"The author has done a fine job of waltzing on the small stage of the layperson's comprehension of lake ecology, all the while making it both an enjoyable and worthwhile read."
—**Rick Mincher**
Former President,
Upper Saranac Lake Association

"Such a comprehensive effort. I appreciate that the author is trying to reach a broad, but curious and informed audience. I think the writing works, and the humor helps…"
—**David Gibson**
Partner,
Adirondack Wild

lakesecrets.com

The Secret Life of a Lake

The Ecology of Northern Lakes and their Stewardship

THE SECRET LIFE OF A LAKE

THE ECOLOGY OF NORTHERN LAKES AND THEIR STEWARDSHIP

Peter Tobiessen

Graphite Press

PUBLISHED BY GRAPHITE PRESS

Copyright © 2012 by Peter Tobiessen

All rights reserved. No part of this book may be used or reproduced without the written permission of the publisher. Graphite Press, Niskayuna, New York.

www.graphitepress.com

Original illustrations by Peter Tobiessen.

Front cover photograph of sunset over Sacandaga Lake:
Copyright © 2006 by Patricia Crane

Back cover photograph of the author:
Copyright © 2011 by Joanne Tobiessen

LIBRARY OF CONGRESS CATALOGING-IN-PUBLICATION DATA

Tobiessen, Peter, 1940–
 The secret life of a lake : the ecology of northern lakes and their stewardship /
Peter Tobiessen.—1st ed.
 p. cm.
 Includes bibliographical references.
 Summary: "A presentation and discussion of the ecological aspects of life
within and around temperate freshwater lakes of the northern United States and Canada"
 —Provided by publisher.

ISBN 978-1-938313-00-4 (alk. paper)

1. Lake ecology. 2. Lake ecology–New York (State) 3. Freshwater habitats.
4. Freshwater habitats–New York (State) 5. Lake conservation. I. Title
QH104.5.N45T63 2012
577.63097–dc23

2012017406

The contents of this publication are for informational purposes only, and not a substitute for professional advice. The author and publisher disclaim any liability with regard to the use of such contents. Readers should consult appropriate scientists or professionals on any matter relating to lakes, the environment, and their management.

FSC
www.fsc.org

MIX
Paper from
responsible sources
FSC® C004071

PRINTED IN CANADA

First Edition

10 9 8 7 6 5 4 3 2 1

"In the end we will conserve only what we love, we will love only what we understand, and we understand only what we are taught."

—Senegalese environmentalist Baba Dioum, from his 1968 address before the meeting of the International Union for the Conservation of Nature, New Delhi.

To Joanne, who inspires me daily, and to our son Josh, who has grown to love Sacandaga Lake almost as much as I do.

Contents

Preface xv

CHAPTER 1
Introduction 1

CHAPTER 2
The Past 5
   Rocks, Faults and Glaciers   5
   Forests   14
   Humans Arrive   16
   Early Settlers   19
   Summer Tourists   20
   Human Impact on the Region   25

CHAPTER 3
Why Water is Weird 29
   The Chemistry of Water   30
   The Water in Lakes   31
   Winter Ice   32
   Seasonal Cycles   36
   Internal Mixing   39
   Light   40

CHAPTER 4
Chemicals in the Water, both Good and Bad 47
   Gasses   48
   Nutrients   54
   Other Chemicals   60
   Trophic Status   62

CHAPTER 5
Some Basic Concepts 65
   Habitats   65
   Basic Niches   67
   Bioaccumulation   69

CHAPTER 6
The Producers I: Algae 73
   Algae   75
   Algal Productivity in a Lake   86

CHAPTER 7
The Producers II: Vascular Plants or "Weeds" 91
   Aquatic Adaptations   92

Vascular Plants   99
Wetlands   103

CHAPTER 8
The Consumers I: Microscopic Beasts & Immature Bugs 107
Protozoa   108
Rotifers   110
Cladocerans   113
Copepods   117
Insects   119

CHAPTER 9
The Consumers II: Fins, Feathers & Fur 127
Fish   128
Birds   135
Mammals   139

CHAPTER 10
Community Interactions 145
Species Competition in Nature   146
Food Webs   150
Bottom-Up Control: The Case of Lake Washington   152
Seasonal Changes   154
Top-Down Control   155
Bioaccumulation, Again   160

CHAPTER 11
Human Impacts 163
Acid Precipitation   163
The Effect of pH on Aquatic Communities   165
Mercury   168
Invasive Plants   171
Invasive Animals   176
Nutrient Enrichment   179
Human Impact on Sacandaga Lake   180

CHAPTER 12
The Future 183
Climate Change   183
Other Concerns   188
Any Good News?   189

APPENDIX I
Measurement Unit Conversions 193
Metric System Conversions   193
Metric to English System Conversions   193
Temperature   194
Quick Estimates   194

APPENDIX II
More About pH 195

APPENDIX III
More About Weird Water 199

APPENDIX IV
Taxonomy & Scientific Classification................................203

APPENDIX V
A Primer on Lake Stewardship 205
    Water Quality 205
    Aquatic Plants 208
    Lakeshore Management 208
    A Few Sources to Assist in Lake Stewardship 209

Glossary...211

Endnotes 223

Acknowledgements 233

Preface

When I was young, my family traveled from eastern Pennsylvania to Wisconsin every few years for a two-week visit with my father's relatives. About one week of that visit was spent in a cabin on an island in a small lake, just north of Milwaukee. Some of my fondest memories as a child came from those short visits—swimming, fishing, and thinking (as Water Rat informed Mole in Kenneth Grahame's *Wind in the Willows*), "There is nothing half so much worth doing as simply messing about in boats." I didn't know it at the time, but a significant imprinting of things aquatic was taking place in my childhood consciousness.

Many years hence, after a graduate degree resulting from four years' study of desert plants, my family arrived in Schenectady and I accepted a position at Union College. However my affinity for lakes remained strong. Fortunately, we were able to buy a small cabin on the shore of Sacandaga Lake, just outside Speculator, New York, in the Adirondack Mountains.

Since there are few deserts in upstate New York, I began looking around for research projects after I arrived at Union. In the mid 1970's, Union College and the Village of Scotia were awarded a grant from the EPA to do an Environmental Impact Study (EIS) on the hydraulic dredging of Collins

Lake in Scotia. At that time, the lake contained dense growths of curly-leaved pondweed—so dense that large portions of the lake were unusable for much of the season. Hydraulic dredging was a relatively new method at that time for controlling aquatic plant growth. A few months into the grant, I was asked to join the project, and soon I was transformed from a desert/terrestrial ecologist into studying aquatic systems. Of course, this was perhaps preordained from my early childhood experiences in Wisconsin.

Since then, I have taught Aquatic Biology at Union for many years, carried out research on Collins and other nearby lakes, worked on a few EIS's at other lakes, and sporadically collected reams of data on Sacandaga Lake. Understanding these data about this specific lake soon became my scientific focus.

With time, I developed a strong interest in the broader community of folks who are fond of lakes. I joined the local lake association that includes Sacandaga Lake, the Lake Pleasant Sacandaga Association (LPSA). I soon realized that members of the LPSA—all intelligent people who love their lakes—knew very little about them. Lakes in the Adirondacks and elsewhere are vulnerable to all sorts of insults, such as pollution, invasive species, and over-development. The more we understand these complex aquatic systems, the better chance we have of avoiding or mitigating such challenges to our lakes. This was my main motivation for writing this book.

I have tried to create a book at a level somewhere between the mostly descriptive genre (limited to describing a lake's inhabitants) and the more rigorous scientific monograph. I wanted to write not only about *who* lives in our lakes, but *why* they look the way they do and *how* they interact with other creatures and their physical environment. However, understanding the "how" and "why" as well as the "who" requires a little background in the basics of the physical aquatic environment and the biology of the organisms in question. Making such discussions readable and accessible to a general readership is an ambitious goal, but I thought that helping the general public understand these marvelous aquatic systems was worth a try. Ultimately, my purpose was to help the curious lake enthusiast better understand the dynamics within these complex ecosystems.

For instance, why do algae have such bizarre shapes? How do small aquatic animals defend themselves against larger predators? How has evolution shaped these organisms? How does the addition of a pollutant affect a lake, and how can we detect a problem before it has a major impact? Why do certain chemicals increase in concentration as they pass from organism to organism within a food web? What effect will invasive plants and animals have on our lakes, and when are controls feasible?

The discussion of such topics sometimes requires the use of scientific terms and names. Their initial use appears in bold italic type, and you can always refer to the glossary for a more extended definition. In addition, since most scientific research uses the metric system of measurement, many of the dimensions, temperatures, and masses are provided in metric units. If you are unfamiliar with this system, or just want a refresher comparing metric with English units, please refer to Appendix I. If a few concepts seem a bit too technical, just skim over them to get the general idea, and then move on. Most of the book should be well within your comfort zone.

Although this book is focused on an Adirondack lake, the principles of lake ecology discussed here are relevant to all lakes that are covered by ice during their seasonal cycle. The more we understand these beautiful, complex ecosystems, the deeper appreciation we will have for our lakes, and the wiser our decisions will be in preserving them.

—P. T.

1

Introduction

Humans have always been drawn to bodies of fresh water. At first it was for the necessities of sustenance and transportation. Later they were attracted by the recreational opportunities of fishing, boating and swimming. However, for many of us there is another dimension to this attraction, one at a more spiritual or primal level. Yeats wrote, "I hear the lake water lapping with low sounds by the shore.../ I hear it in the deep heart's core."[1] It is also not surprising that Thoreau, when he was searching for inspiration in his struggle to resolve the conflict of nature and civilization, chose a site next to a lake—Walden Pond—to build his cabin. He wrote:

> "A lake is the landscape's most beautiful and expressive feature. It is the earth's eye; looking into which the beholder measures the depth of his own nature."[2]

Some have even suggested that our affinity for lakes arose from our prenatal sloshing in maternal amniotic fluid. I only know that when I am near a lake, I feel a kind of peace that I rarely find elsewhere.

However, below the peaceful surface of a lake is a dynamic and complex ecosystem of fascinating creatures interacting in weird and wonderful ways. The typical person enjoying a canoe or kayak trip on a lake is totally unaware of what lies beneath the lake's surface—in essence the lake's "secret life." For instance, the curious observer of a lake may ask: What determines the shape of a lake? Why does a lake freeze from the top down instead of the bottom up? How can an apparently homogeneous volume of water in a lake support hundreds of species of plants and animals? How do these species interact? What lies in the future for our lakes?

In this book, I will attempt to reveal some of these secrets by explaining the workings of a typical temperate lake—a lake that goes through a predictable thermal cycle including a period of ice cover. Although the focus will be on a lake in the Adirondacks in northern New York and other nearby lakes, the principles discussed will be relevant to most medium-sized lakes in the northern United States and Canada. To make the discussion concrete, I will use as a model system Sacandaga Lake, located near the villages of Speculator and Lake Pleasant in the town of Lake Pleasant in Hamilton County, New York (see Figure 1.1). You can see an aerial photograph of this lake region in Figure 1.2.

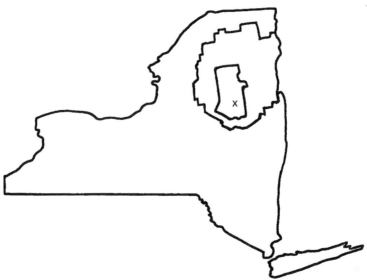

Figure 1.1
New York State, with Adirondack Park shown in the northeast corner. Totally inside the park is Hamilton County, and Sacandaga Lake is indicated by the "X" inside the county.

Figure 1.2
An aerial photograph showing Sacandaga Lake (center), Lake Pleasant (right) and Fawn Lake (left).
Original Image Copyright © 2012 by DigitalGlobe

At the outset, for those readers familiar with upstate New York, I must distinguish Sacandaga Lake, a natural lake at the headwaters of the Sacandaga River, from the more widely known Great Sacandaga Lake, a large reservoir about 25 miles downstream. The latter was formed in 1930 by a dam on the Sacandaga River at Conklingville to control floods downstream on the Hudson River and to provide additional water to the Hudson during periods of low flow. Whereas the largest dimension of Sacandaga Lake is only about 2 miles, Great Sacandaga Lake is indeed great. It is 29 miles long with a shoreline of 125 miles, about the same size as Lake George.

Sacandaga Lake is a beautiful lake, and it is similar to many of the estimated 2,000-plus lakes in the Adirondacks. I chose it as my model because I have spent parts of my summers for over three decades in a small cabin on its shore. About 40% of the shoreline contains private (mostly summer) residences, and the remaining 60% is controlled by the New York State Department of Environmental Conservation (NYSDEC). Half of the latter makes up Moffitt's Beach State Campground with about 260 campsites.

The rest of the NYSDEC land is designated by the Adirondack Park Agency as "wild forest," one of the more restrictive land use classifications regarding human development. Just downstream from Sacandaga Lake is Lake Pleasant, and about a quarter mile to the northwest is Fawn Lake. Since all three lakes will be mentioned in this book, I've listed their "vital statistics" in Table 1.1. Because Fawn Lake has no human shoreline development, it will be compared with Sacandaga Lake as a "control," or what we might have expected Sacandaga Lake to be like before the arrival of humans.

Table 1.1
Descriptions of Three Lakes in the Adirondack Region.[3]

	Sacandaga Lake	Lake Pleasant	Fawn Lake
Elevation Above Sea Level (ft)	1726	1725	1701
Surface Area (acres)	1620	1547	289
Maximum Depth (feet)	73	69	62
Average Depth (feet)	28	29	34
Percent Shoreline Development	± 40	± 70	0

These lakes are located in Hamilton County, the fifth largest county in New York State and its least densely populated. It has an area of about 1,800 square miles, 4.8% of which is water. It is nearly the size of the state of Delaware, yet its 2010 census records only 4,836 permanent residents. During the summer, its population increases by a factor of five. As of 2012, there were no permanent traffic lights in the entire county. In other words, this is a very rural area, and the lakes' watersheds are mainly forested.

2

THE PAST

To understand a body of water in its current condition, we must study its past. In the case of Sacandaga Lake, that means we must go back more than a billion years, when the bedrock that underlies most of the Adirondacks was formed. It was also in the distant past when geological events defined the shape of Sacandaga Lake as well as most of the other lakes in the Adirondacks. But before we can understand these events, we need to take a detour into a more recent episode in the history of science.

Rocks, Faults and Glaciers

Plant and animal geographers had long puzzled over how the existing ranges of similar organisms occurred on different continents. For instance, close relatives of plants and animals in North America also occur across the ocean in Europe. In addition, animal groups in South America are more similar to those in Africa than other continents. And then there are those problematic fossils in Antarctica. Land bridges between continents were suggested as transport routes, as well as debris rafts cruising the currents bringing organisms across the oceans. In 1921, Alfred Wegener noticed that the hump on the east side of South America could fit nicely into the

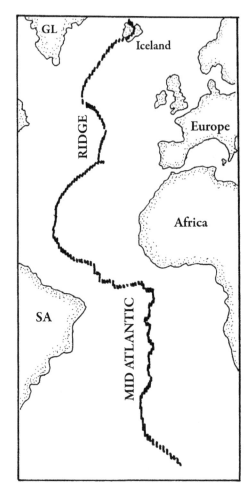

Figure 2.1
The position of the Mid-Atlantic Ridge relative to the continents. Note the location of Iceland at the northern end. GL = Greenland, and SA = South America.

indentation in western Africa, and he came up with the revolutionary idea that all the continents were at one time stuck together. He even suggested a name for this supercontinent—Pangaea, Greek for "all lands." However, he had no idea about the mechanism for this *continental drift*. His first hypotheses for this mechanism were easily discredited, and eventually his ideas were dismissed by the scientific community. We all knew that *terra firma* was, well, firm. He died in 1930 with his ideas at best in limbo and at worst ridiculed.

Subsequent research in oceanography eventually provided data to support Wegener's hypothesis. Oceanographers knew that there was a region in the middle of the Atlantic Ocean called the Mid-Atlantic Ridge (MAR), which was shallower than the ocean floor to the west and east, and it contained distinct seismic activity. Iceland, with its volcanoes and hot springs, is a hot spot on the northern extreme of this ridge, where it has risen so high that it is above sea level (Figure 2.1). Could this ridge be part of the mechanism that causes the continents to move?

Sophisticated coring techniques of the deep ocean sediments eventually showed that the seabed nearest the MAR was very young, and it grew sequentially older as it progressed outward, both to the east and west. As new seafloor is formed at the MAR, it spreads out to the east and west, pushing the continents of Europe and Africa away from North America and South America (Figure 2.2). In fact, as you read this, the Adirondacks are fleeing England at a rate of almost 4 cm (about 1.5 in) each year.

Plate Tectonics

So the continents are moving, floating on top of the earth's molten mantle layer on large plates, their motion powered by convection currents within the Earth's mantle. The plates are crashing into one another and then separating, and they have been doing so for billions of years. As these plates collide, mountain ranges form, and volcanoes emerge where one plate slides beneath another. When plates slide past each other, large faults develop generating earthquakes, such as those along California's San Andreas Fault. This field of geomorphology is known as *plate tectonics*, and it is as fundamental to the geosciences as evolution is to biology.

It is too bad that Alfred Wegener did not live to see his idea become accepted by the scientific community, but that is how science proceeds. Ideas arise, they are tested and debated, and if supported by subsequent studies they are generally accepted. If succeeding studies do not support them, they are relegated to obscurity. However, sometimes new methodologies develop to retest previously rejected hypotheses, and ultimately they gain credibility. Such was the case with plate tectonics. I can only hope that Wegener's spirit is looking down on us from above to see his original idea become one of the major unifying concepts of geology, and he is pumping his fist and shouting an ethereal "Yesss!"

I have delved into the topic of plate tectonics because it explains much

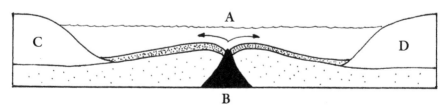

Figure 2.2
A section through the Mid-Atlantic Ridge (A) showing the magma (B) rising to the surface and the ocean floor (dark stippled) spreading to force the two continents, North America (C) and Europe (D) apart.

of the geology of the Adirondacks as well as some specific characteristics of Sacandaga Lake. A comprehensive geological history of the Adirondacks would require volumes, so I will limit my discussion only to topics directly relevant to our lake.

Colliding Continents

When continents collide, they often form mountain ranges in a process called *orogeny*. This is currently occurring in the Himalayas because the Indian subcontinent crashed into the Eurasian Plate a mere 50 million years ago. Because the collision continues to this day, the Himalayas are still rising. Similar collisions have occurred in the past on the east coast of what is now North America. Approximately 1.3 to 1.1 billion years ago, one collision called the Grenville Orogeny produced a huge mountain range, thought to be on the same scale as the current Himalayas. It extended over a wide range from eastern Canada down to Alabama and Mississippi, including the current Adirondacks. Older rocks lying under this mountain range, about 18 mi deep, had been deposited by ancient seas, and they were contorted and heated by the pressure of the overlying mountains to become molten and metamorphosed. Layers of limestone were changed into marble, and molten magma intrusions spread throughout these layers,

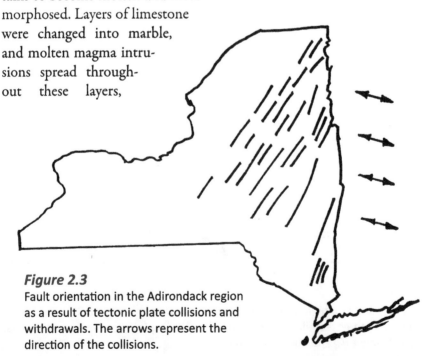

Figure 2.3
Fault orientation in the Adirondack region as a result of tectonic plate collisions and withdrawals. The arrows represent the direction of the collisions.

The Secret Life of a Lake

solidifying into igneous rocks.[1] During the next 400 million years these mountains were eroded away to form a flat plain, but the underlying older igneous and metamorphosed layers were left in place. Between 650 and 550 million years ago, the Adirondack region was stretched in the direction of the current Massachusetts/Connecticut region, leaving major faults oriented perpendicular to the stretch direction, extending north-northeast to south-southwest (Figure 2.3).[2]

Subsequent comings and goings of continental plates

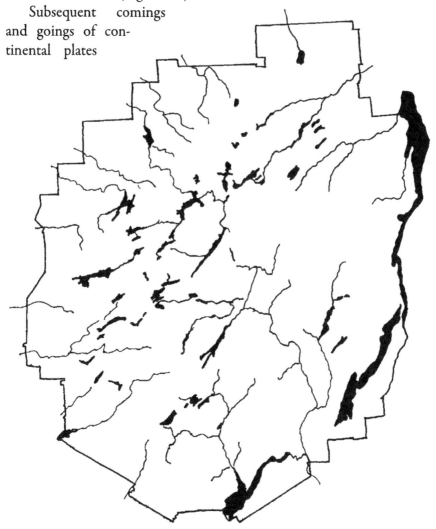

Figure 2.4
A diagram of the Adirondack region showing the orientations of the major lakes and drainage patterns. Note the similarity to the fault orientations in Figure 2.3.

also came from this direction, so resulting faults and foldings are similarly oriented. This is important to Sacandaga and other Adirondack lakes because faults are weakened regions in the earth's crust, and erosion is accelerated along them. Erosion results in depressions, which eventually fill with water to become lakes many years later. A view from high altitude shows that most lakes in the Adirondacks are long and thin and are oriented north-northeast to south-southwest, a result of these ancient faults (Figure 2.4). Sacandaga Lake is unusual in this regard because it is formed by a few small adjacent faults, with the area between also filled with water, giving the impression of a non-linear lake. In fact, its name during the 1800's was Round Lake.

About 500 million years ago, the Adirondack region was covered by a shallow sea, which deposited sandstones and calcium-rich limestone sediments over the area. However, subsequent erosion stripped away most of these sedimentary deposits, leaving them only where they were protected in deep faults. The presence or absence of limestone and its metamorphosed relative, marble, in a lake's watershed is critical to understanding the lake's resistance to acid deposition, because limestone is made of calcium carbonate that can neutralize acids. This relationship is further discussed in Appendix II, as well as Chapters 4 and 11.

The current New England area was affected by later orogenies—the Taconic around 450 million years ago, the Acadian around 400 million years ago and the Alleghanian around 320 million years ago. However, these had relatively little effect on the Adirondacks, only producing a few more faults mostly oriented in roughly the same direction as previous ones. However, they did form the Taconic Mountains north of New York City and much of the Appalachian Mountain chain. Sometime before 200 million years ago, all the continental plates were connected, forming Wegener's supercontinent Pangaea. He was right about this, and he even estimated the correct time that it formed. However, around 200 million years ago Pangaea began to break apart, and the eventual result of this dispersion is the current position of the continents, with the Atlantic Ocean still getting wider as you read this.[3]

During this turbulent period of mountain building described above, the Adirondacks were relatively quiescent. They were a flat plain of ancient rocks, overlain by a layer of sedimentary rocks deposited by the shallow sea hundreds of millions of years before. However, currently the Adirondacks

The Secret Life of a Lake

are rising in elevation at a rate of two to three millimeters each year (a millimeter is about the thickness of a dime), a rapid rate for such a geological process. Subsequent erosion from this rising land mass removed most of the remaining sedimentary deposits, leaving them only in certain localized areas protected by faults.[4] Currently the Adirondacks are made up of a rising, eroding mass of ancient igneous and metamorphosed rocks developed during the Grenville Orogeny, silica rich and carbonate poor, that are poorly equipped to buffer the acids generated by decaying organic matter or deposited through atmospheric pollution.

The Impact of Glaciers

Around 1.8 million years ago the Pleistocene epoch began, which is characterized by a series of global glaciations. There were many major glacial advances during the Pleistocene, and the most recent, called the Wisconsin, began about 70,000 years ago and ended about 10,000 years ago. At its maximum, the Wisconsin ice front extended down to Long Island, across the top of Pennsylvania, and continued through the northern midwest (Figure 2.5). Although there were many previous glaciations during the Pleistocene, the Wisconsin extended farther south in the eastern U.S. so that nearly all the evidence of the earlier glaciations was obliterated or

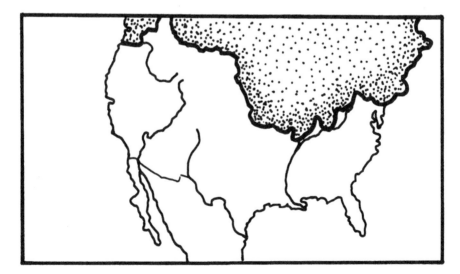

Figure 2.5
The maximum advance of the Wisconsin glaciation about 13,000 yrs ago. Note that the Adirondacks were totally covered.

severely modified. Ice thickness was estimated to average about one mile in central New York, depending on topography, so it had a significant effect on the Adirondacks. Mountains were rounded off and their tops ground down. Valleys were carved out further, and all the surficial features such as loose rocks and soil were scoured out and transported southward by the advancing ice sheet. The glacier stopped advancing at the present position of Long Island, which owes its existence to the debris pile deposited there by the retreating glacier. Such aggregates of rubble collected at the front of an advancing glacier, then deposited when the glacier retreats, are called **terminal moraines**. As the glacier retreated, there were intermittent shorter cyclic periods of advance and retreat, so that smaller terminal moraines were formed.

Lakes can be formed in the depressions created by glaciers, as well as bounded by the terminal and lateral moraines, though these are more common in the mountainous west than the Adirondacks. A retreating glacier can also break off large blocks of ice that can then be surrounded by layers of rubble washed down from their receding parent. When the ice block melts, it leaves a depression in the rubble that is roughly circular and has a broad, flat bottom, similar to the shape of a kettle. Resultant lakes in these depressions are called **kettle lakes**, and Round Lake to the south of the Adirondacks is a good example. The northern Midwest has many kettle lakes, but they are not so common in the Adirondacks. However, some have suggested that the broad, flat-bottomed basin in the western portion of Sacandaga Lake may be at least partly due to kettle formation, although bedrock outcrops along the shore do not support this explanation.

The tremendous weight of the glacier depressed the rock strata of the Adirondack region, and as the glacier retreated, the strata began to rebound. In fact, it is still rebounding (geological processes can be very slow!), and some geologists think that it is this rebound that is causing the Adirondack Mountains to continue to rise in elevation.[5]

The glacier was a huge benefactor to those of us who love lakes. As a lake ages, it naturally tends to fill in with debris such as silt and organic material washed in from tributaries. In addition, photosynthesis by plants and algae within the lake produces organic material that also settles to the lake bottom. All this material accumulates, decreasing the lake's depth and eventually filling in shallower lakes over time. Over the millennia, lakes are transformed into marshes, then into meadows and forests. This is a natural process that occurs with all lakes. The glacier scoured all these deposits from the depressions that were formerly lakes—in most cases right down to the bedrock. In other words it made new lakes where they had not existed for thousands of years. It is no accident that the 2,000-plus lakes in the Adirondacks and the presumed 10,000 lakes in Minnesota are located in recently glaciated areas. Alabama and Oklahoma do not have thousands of natural lakes. The glacier also produced the Great Lakes, the Finger Lakes and most of the lakes in Canada. So the next time you look out over a beautiful Adirondack lake, you can thank global cooling.

You may wonder how these massive glaciers formed and why they advance and contract in such a marked cyclical fashion. The answer is that glaciers form and melt by a process that is governed by a positive feedback loop. *Positive feedback* occurs if the result of the process (ice, in this case) accelerates the formation of that result (the freezing of more ice).

The Earth's energy budget depends on the amount of energy entering the Earth's system, mostly from the sun, relative to the amount of energy leaving it through reflectivity, long wave radiation and a few other pathways. If there is a balance, the earth remains at a constant temperature. Suppose for some reason the Earth cools a bit, resulting in an increase of snow and ice cover. Both snow and ice have a higher reflectivity than bare ground, vegetation or water. The result will be that the Earth's reflectivity will increase, more sunlight will be reflected away from the Earth into space and the Earth will cool a little. This leads to the formation of more ice and snow, more reflectivity, and a resultant increase in ice and snow.

This positive feedback loop will ensure that, once a glacier starts to advance, it will continue to do so until it reaches the lower latitudes with more direct sunlight and temperatures warm enough to reverse this process. Once the glacier starts to retreat, more bare ground is exposed that absorbs rather than reflects sunlight, and the Earth will warm a little more. This will result in more ice and snow melt. Cyclic changes in the relationship of the Earth's orbit around the sun, as well as cyclic wobbles in the Earth's axis, are thought to initiate glacial cycles, but the positive feedback explained above is critical in amplifying these effects.

Positive feedback systems are inherently unstable, so it is important that we can distinguish between positive and negative feedback loops when we consider the impact we humans are having on our planet. We are currently in a period of warming, with glacier and icepack melt on a global scale. This is directly analogous to the glacial retreat model described above with its positive feedback. In addition, as will be discussed further in Chapter 12, the warmer the Earth's surface becomes, the more water is evaporated into the atmosphere. Water is a greenhouse gas, so that will tend to make the Earth warm up even more rapidly, another positive feedback loop. So the more the Earth warms, the more it will tend to warm in the future unless something is done to reverse this process. It is these feedback loops, both positive and negative, that make future climate predictions so difficult.

Forests

As one travels north from New York State through Canada to the northern reaches of the continent, the vegetation changes in predictable and recognizable ways. In the far north, the cold temperatures, short growing seasons and mostly frozen soil allow only herbs and shrubs to survive. The few tree species are dwarfed and shrub-like. This type of vegetation is known as ***tundra***. Just to the south of the tundra margin is the vast ***boreal forest***, dominated by the conifers larch, jack pine, balsam fir and spruce (both white and black), as well as the deciduous aspen and paper birch. In colder climates, the evergreen habit of most conifers allows a longer period of photosynthesis than deciduous species, so conifers tend to be more common at higher latitudes. The deciduous aspen has photosynthetic bark, which allows photosynthesis even when the tree is leafless. (In fact, the next time you see an aspen, scrape the thin, grayish bark off a twig and see the bright green photosynthetic tissue underneath.) As we travel farther south, the proportion of deciduous trees increases at the expense of evergreens. Deciduous sugar and red maples, beech, yellow birch and ashes, among others, become more prominent as well as the evergreen hemlock, to form the typical ***northern hardwood forests*** we see in the lower elevations of the Adirondacks.[6]

As the glacier advanced, it pushed these vegetation zones southward into refuges in the Mid-Atlantic States and onto the Appalachian ridges of the Southeast. When the glacier retreated, the vegetation followed it back north. If we had stood on the current shore of Sacandaga Lake for the last 10,000 years, the vegetation sequence passing us would be similar to what we found on our trip from the tundra to the northern hardwood forest described above. How do we know? Higher plants use pollen for reproduction, and many plants, especially trees, are wind pollinated. Most of the pollen wafting about in the breezes eventually falls to the ground, or in the case of a lake, it settles to the bottom of the lake. The cell wall around a pollen grain is very resistant to decay and can remain virtually unchanged at the bottom of a lake for thousands of years. Over time the sediment builds up on the lake bottom like a layer cake, with the oldest layers on the bottom and the more recently deposited layers toward the top. By taking a core of this sediment, we have a sequential history of the vegetation in the vicinity of the lake after the glacier cleaned out the previous sediment layers and dropped them off in Long Island. Radiocarbon dating of organic deposits at various positions in the core calibrate the position of the pollen to actual dates in the past.

A core from a lake-turned-bog within 40 miles of Sacandaga Lake and another from Heart Lake farther north show this progression nicely. From these and other Adirondack cores, researchers found mainly herbaceous and shrub species before 10,500 years ago, and mainly spruce, fir, aspen and larch shortly thereafter. Between 9,600 and 7,050 years ago came pollen from jack pine first and later white pine, as well as birch and oak. Hemlock appeared between 7,050 and 4,650 years ago, in addition to the deciduous species maple and beech. The pollen in samples younger than 4,650 years ago reflects the current forests in the area, so the northward vegetation migration was essentially complete in about 6,000 years. In the last 4,000 years or so, there have been minor warmings and coolings that are also reflected in the pollen profile. Approximately 125 years ago, ragweed pollen appeared in the sediment indicating significant forest clearing, most likely by European settlers. Although this summary has been grossly simplified to touch on just the high points, it does give an overview of the process.[7,8]

As the vegetation surrounding the lakes changed, so did the water chemistry within the lake. In the Heart Lake study, the authors studied the chemistry of sediments and the assemblages of algae in the core. They found that as the vegetation surrounding the lake changed with time, so did the lake's water acidity and nutrient levels. These fluctuations resulted in changes to the biological community within the lake.[9]

I mention the effect of the surrounding vegetation on the chemistry and resulting biology of a lake, since modern shoreline development and major watershed disruptions from fires and logging have inevitably affected water quality in Sacandaga, and will continue to do so in the future. Also, we must be able to separate the natural effects of forest change, such as forest development after fires, from human-induced impacts, such as atmospheric acid deposition. For instance, the increase in acidity found in Adirondack lakes during the 1960's could have been the result of increasing levels of organic matter deposited in the soils of watersheds as the forests recovered from the extensive deforestation and forest fires that occurred around 1900. Only subsequent research clarified the link between increases in lake acidity and acid rain, as discussed further in Chapter 11.

The temperature change of the Earth from the last glacial maximum to the current interglacial period is estimated to be about 10.8 °F (6 °C) over a period of 6,000 years, when the vegetation zones reached their current positions. However, the Earth's climate is warming again. Some estimates of temperature change from recent global warming models suggest a change of about 5.4 °F (3 °C) over a period of only 100 years.[10] This recent rate projection is about 30 times faster than the rate found after the glaciers retreated. So plants and their associated biological communities will be on the march northward again, this time in a more rapidly changing thermal environment. For rare plant species with limited ranges, there is some question as to whether they will be able to migrate rapidly enough to avoid extinction. In addition, since humans have fragmented natural habitats with intervening agricultural, urban or suburban development, migration northward will be impeded and many species will no doubt be lost.

Humans Arrive

It is impossible to be certain when the earliest humans arrived in New York State, since the glacier scoured away all traces of their presence before 10,000 years ago. However, the earliest artifacts of Paleo-Indians in New York State are fluted spear or javelin points found sporadically along the Hudson Valley. They are called *Clovis points* and they have been dated to approximately 7,000 years ago.[11] No doubt these Paleo-Indians spent most of their time in the warmer valleys like the later native groups, but they did come into the Adirondacks, most likely to gather game and fish. Indeed, a Clovis point was found in the Town of Wells, only a few miles south of Sacandaga Lake, suggesting that the Sacandaga area has had human visitors for at least 7,000 years (Figure 2.6).

Subsequent native groups have no doubt traversed the trails to Sacandaga Lake, as indicated by the large number of arrowheads and other native artifacts found around Sacandaga and adjacent lakes. But the group

Figure 2.6
A Clovis point found in the Town of Wells, indicating native habitation in the area near Sacandaga Lake 7,000 yrs ago.
From the collection of the Historical Society of Lake Pleasant and Speculator.

we know the most about during recorded times is the Mohawks, one of the tribes of the Iroquois Nation. The Mohawks were the eastern-most tribe of the Iroquois, and they laid claim to most of the Adirondacks as their tribal hunting grounds. In fact, it was this territoriality that gave the Adirondacks their name. An Algonquin tribe from the north called the Montagnais also hunted in the Adirondacks, and whenever hunting parties of the two tribes met, there was conflict. The Mohawks were relatively prosperous, since they farmed in the fertile Mohawk Valley, but the Montagnais were hunter-gatherers. When food was scarce, they resorted to peeling the bark from twigs and eating the soft inner layers. The Mohawks derisively called them "ratirontaks" meaning "tree eaters" or "bark eaters" in the Mohawk language, which sounded like "Adirondacks" to European settlers.[12] In 1838, Professor Ebenezer Emmons of the New York Geological Survey first christened the Adirondacks with the derisive Mohawk name for the Montagnais.[13]

It is generally thought that the early native tribes did not have permanent settlements in the Adirondacks, but used them only for game and fish. Hamilton County historians Aber and King state that only two native families—those of Peter Sabattis at Long Lake and Sabael Benedict at Indian Lake—had residences in the area before 1800. Benedict arrived soon after the Revolutionary War and is considered the first permanent resident of Hamilton County. Neither Benedict nor Sabattis were Mohawks but Abenakis of the Algonquin lineage.[14]

Around 1800, an old Mohawk named Captain Gill and his family lived on the shores of Lake Pleasant, adjacent to Sacandaga Lake. European settlers had begun to arrive, and by 1806 about a dozen families lived in the area. Captain Gill would often be their guide. He was apparently quite a story teller, and he would narrate to the newcomers some age-old native legends. One involved Sacandaga Lake.[15]

He told of a long-forgotten tribe from this area that was suffering from an especially hard winter. There was much starvation because the vegetation was killed to the roots, and game had migrated to warmer valleys. At a council fire, there was heated discussion between the elders and some younger braves who wanted to leave the area to look for a more hospitable site. The elders wanted to stay and held firm, stating that the famine was brought on by the "Master of Life" for their sins, and that after a period of suffering, the Master would renew their bounty. To flee would bring the Master's wrath. The young men were not convinced, and they rose in anger and killed the elders. In remorse, the young rebels decided to sanctify their elders by decapitating the corpses, burning the bodies and sinking their heads ceremoniously in Sacandaga Lake. The leader of the rebels led the canoe procession, and, as he fashioned a large weight to the heads to sink them, his canoe was swamped and he was dragged to his death by the weight. The following week, bubbles and other signs of disturbance appeared on the lake, and on the sixth day a monstrous head rose from the water. By the seventh day, giant bat-wings with claws had grown from the head, and it flew out of the lake to torment the murderers. Wherever they went, the head pursued them. Some say it turned them to stone, others said it pursued them onto the prairies, and still others said the head tortured them by keeping them forever young so that they would not be able to end their suffering.

We don't know whether the good Captain Gill spun these yarns to keep the settlers on edge or whether there was a kernel of truth in them, as is the case with many old legends. (A hill just south of Lake Pleasant is named "Indian Head.") If the latter were the case, there may have been a long-standing native settlement in the area before the late 18th century. Even if that were true, the native impact on the area surrounding Sacandaga Lake was most likely minimal. But the legend remains, and on those foggy mornings when we look out onto Sacandaga Lake and hear mysterious sounds—is it a loon or…?

The Secret Life of a Lake

Early Settlers

Access to the area was enhanced by the War of 1812. Our young nation was confronted by the British foe in Canada, and Congress deemed it critical to develop three roads leading northwards to Canada. They were called "military roads" and one passed from Northville, through Wells, north of Lake Pleasant and Sacandaga Lake (called Round Lake at this time), and on to Raquette Lake and Saint Lawrence County. It was completed to the St. Lawrence Turnpike in 1815.[16] By now there was a small settlement near the northern shore of Lake Pleasant named Newton's Corners after an early settler and storekeeper. On the southern shore was another small settlement called "Lake Pleasant." The Military Road from Newton's Corners to Wells was maintained and eventually improved, but the road north of Newton's Corners fell into disuse (Figure 2.7).

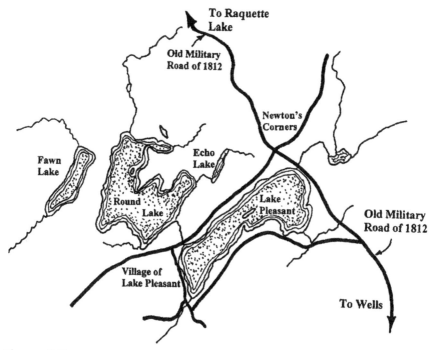

Figure 2.7
A modified diagram of a map of the area, drawn from 1821 data.[14] Note the Old Military Road from Wells, heading north toward Raquette Lake and the St. Lawrence. The crossroads to the north of Lake Pleasant is Newton's Corners (currently Speculator), and the current Sacandaga Lake was called Round Lake at that time.

Table 2.1
The population of the Town of Lake Pleasant from 1820 through 1950, from the Census of the State of New York.

Date	Population	Date	Population
1820	313	1920	323
1840	296	1930	540
1860	356	1940	584
1880	343	1950	696
1900	469		

By 1812, the area around the lakes had a population of about 200, and the Town of Lake Pleasant was established, comprised of 198 square miles or about 10% of the area of Hamilton County.[17] The Hamlet of Lake Pleasant is a small settlement within the town. Throughout the 1800's, the population of the town showed little change, but there was a slow increase starting around 1900 (Table 2.1).

The area around the lakes did have some agriculture, as shown by the state census of 1865, but little of the land was plowed (Table 2.2). These data show that the town was never heavily agricultural. A 1913 forestry map shows that virtually all this "improved" land surrounded Lake Pleasant, and was not in the Sacandaga Lake watershed, a fact that will be recalled later in Chapter 11 when discussing the effect of human development on the biology of our lakes.

Summer Tourists

Just after 1900, the area saw a boom in tourist traffic. Many large hotels sprang up because the region provided a sense of wilderness that was relatively accessible to major population centers. According to a 1906 travel brochure, travelers could catch the night boat from New York City at 6:00 PM and be in Albany by 7:00 AM the next day.[18] Then they could hop on the 7:30 AM train to Fonda, change trains and be in Northville by 10:30 AM. Then they would get on a four-horse "Tally-Ho" stagecoach (Figure 2.8), head up the "improved" military road to Wells for midday dinner, change horses to a six horse team and go the last 10 miles uphill to Speculator for a 7:00 PM supper at the Sturges Hotel. So within a 25-hour span, they could leave stifling, bustling New York City and be in the peaceful wilderness

The Secret Life of a Lake

Table 2.2
Land use categories in the Town of Lake Pleasant in 1865, from the Census of the State of New York.

	Acres	Development
Unimproved Land	38,527	Forested
Improved Land	2,570	Cleared
Plowed	174	
Pasture	923	
Meadows & Grassland	1902	

with reasonably comfortable hotel accommodations. Other areas of the Adirondacks provided a similar summer respite from the sweltering cities to the south.

The trip was not without hardships, however. The route from Wells past Gilman Lake to Speculator was mostly uphill on a muddy, sometimes

Figure 2.8
The "Tally-Ho" stage linked Newton's Corners and Wells in the early 1900's.
From the collection of the Historical Society of Lake Pleasant and Speculator.

corduroy road. Men and children often had to walk to lighten the load for the horses. But compare this well-traveled route to a trip in 1891 by a school teacher on a buckboard. She traveled from Newton's Corners north to Indian Lake, a trip of only 25 miles that took an entire summer day, dawn to twilight. Only then can we realize that Newton's Corners was essentially the end of the road for all but the most intrepid travelers.[19]

In the early 1900's, the round trip from New York City to Speculator using the Hudson River boat from NYC to Albany cost $8.80. Taking the train all the way from NYC cost $11.25.[20,21] Comparing that price with the cost of a room at one of the hotels in the area—$2 to $3 per night or $10 to $15 per week—suggests that either transportation was very inexpensive at that time or the area hotels were pricey. The fact that the average family income around 1900 was $500 to $1,000 per year suggests the latter.

The 1906 travel brochure from the Fonda, Johnstown and Gloversville Railroad Company states that the Sacandaga River Valley resorts in the Lake Pleasant area ensured "an invigorating atmosphere and a salubrious climate," and that "The stage ride is one of the most delightful features of the trip to any of the Sacandaga Valley resorts in the Lake Pleasant region."[22] The brochure somehow forgot to mention the muddy trek up the hill from Wells for some travelers. The brochure also lists the following hotels in the Speculator/Lake Pleasant region (Table 2.3).

That is a total accommodations for 740 guests in this relatively small area. Most came from NYC and many stayed the entire summer. Morley's was described as "one of the finest summer resort hotels in the Adirondacks," and the Osborne Inn had "baths and toilets on each floor" and a "table beyond criticism" (Figure 2.9). So the area was quite a tourist destination around the turn of the century. In 1915, the summer population of Hamilton County was 25,000, whereas the resident population reached

Table 2.3
Hotel accommodations in the Speculator/Lake Pleasant area in 1906.

Hotel	Guests	Hotel	Guests
Osborne Inn	80	Morley's	250
Sturges House	100	Dunham Cottage	30
Lake Pleasant House	75	Lily Lake House	100
Speculator House	25	Fish Mountain House	80

Figure 2.9
Hotels in the Speculator area around 1900: Hamilton County Inn (top) and the Osborne Inn (bottom).

From the collection of the Historical Society of Lake Pleasant and Speculator.

only 4,300.[23]

Around the end of the 19th century the names of the two settlements in the area changed, as well as the name of Round Lake. The Hamlet of Lake Pleasant on the southern shore of Lake Pleasant was well established, so that it had become the county seat of Hamilton County by 1838. But in 1844 Hezekiah Sage, the proprietor of the Lake House hotel, changed its name to Sageville. In 1896 when Sage died, the name of the village reverted back to Lake Pleasant. Meanwhile, the federal Post Office Department sent out a dictum that the post office in Newton's Corners change its name, so in 1896 Newton's Corners became Speculator, named after nearby Speculator Mountain.[24] A local hotel owner wanted the name change, since he thought "Speculator" would sound more attractive to potential tourists than "Newton's Corners." Around this time the State Forestry Commission rechristened Round Lake as Sacandaga Lake, which appeared on maps after 1900.[25] Perhaps there were already too many "Round Lakes" in New York, or perhaps someone realized that it wasn't really round, or perhaps the locals preferred the more tourist-attractive Mohawk name, but the lake is clearly the headwater of the Sacandaga River, a name that was well established during the 1800's. The river's name has been variously defined from the Mohawk language as "flowing grass"[26] or "drowned lands,"[27] most likely referring to the extensive marshlands found along the river in an area now flooded by the current Great Sacandaga Lake.

The tourist heyday of the Speculator area came in the 1920's. William Osborne, of the Osborne Inn family, was a Marine in the First World War, and one of his fellow marines was named Gene Tunney. When Osborne found out that Tunney was a prizefighter, he suggested that Tunney come up to the Osborne Inn to train. He arrived in 1926 to train for the Dempsey fight, and after he won, he returned to train for subsequent fights.[28] Max Baer, Max Schmeling and Maxie Rosenbloom also trained in Speculator, so the area became a Mecca for fight fans. Demonstration fights drew large crowds. Special trains came up from New York City with patrons who watched the fights. Automobile traffic was substantial enough for the Village to have to establish a speed limit in town of 30 mph in 1922.[29]

The depression of the 1930's and the Second World War inflicted a double whammy on the area as a tourist destination. Money was scarce and gasoline was rationed, and occupancy in the area's hotels plummeted. One by one the major hotels were abandoned, or more frequently, burnt to the ground. The Osborne Inn was the last to go, and it lasted until 1969.

Human Impact on the Region

The trip from Wells to Speculator around the turn of the century was difficult for freight wagons, so hotels in the Speculator region had to be largely self-sufficient. There was some agriculture in the area to support hungry tourists, but the local fish and game took a real beating. Most of the hotels had guides to take their patrons out hunting and fishing, and their success was well known (Figure 2.10). No doubt some game was used in the hotels, but the local fishery took the brunt of the harvest. The Sturges House hotel in Speculator served fish on Fridays, and the local lakes provided the main course:

> Each Thursday, the wagon would leave the hotel for the lakeshore with an ample supply of empty nets. In a few short hours, it would come lumbering back up the road with sufficient lake trout to feed from 150 to 200 guests the following day.[30]

Figure 2.10
Adirondack guides and their clients. The Sacandaga Lake/Lake Pleasant area was popular for guided hunting and fishing trips. Note the amount of game for this one party.

From the collection of the Historical Society of Lake Pleasant and Speculator.

Paul Wilbur, the Lake Pleasant Town Historian, recalls his father telling of supplying large numbers of lake trout to the local hotels in the early 1900's, and he was not alone (Figure 2.11). Fish became harder to catch, and as early as 1887 a state-run hatchery was established on Hatchery Brook, a major tributary to Round/Sacandaga Lake. It functioned for 15 years.[31] In 1919 the state closed many streams in the area to fishing, and in 1921 more streams were closed. "Trout fishing was at a low ebb" in the 1920's.[32]

Sustainable populations of lake trout are currently not present in either Lake Pleasant or Sacandaga Lake, and historic over-fishing no doubt contributed to their demise. In nearby Fawn Lake, which is not accessible by road and has a shoreline free of human development, there remains a well established, apparently self-sustaining population of lake trout. More about the demise of the lake trout populations in Sacandaga Lake and Lake Pleasant is covered in later chapters.

Agricultural land use in the area was concentrated around Lake Pleasant, and around 1850 the lake had quite a cultivated aspect, so that a tour-

Figure 2.11
An Adirondack fisherman with his catch, circa 1900.
From the collection of the Historical Society of Lake Pleasant and Speculator.

ist brochure at that time stated, "a number of beautiful grazing farms are opened to view...down to the water's edge." By the turn of the century most of the forests around Lake Pleasant had been turned into fields, so that another travel brochure stated, "Although Lake Pleasant can boast of little wilderness, as clearings nearly encircle it..."[33] The state Fire Protection Map of 1916 confirms these observations, but it also shows that there was very little clearing around Sacandaga Lake.[34] Its watershed remained almost entirely forested. As for Lake Pleasant, agriculture decreased in the area due to better transportation and cheaper food produced elsewhere, so forests have mostly reclaimed the southeastern shore in the later 1900's. The area now appears to be much more of a "wilderness" than it did in 1900.

The forested areas within the watershed also felt the impact of European settlement. Forest fires ravaged the Adirondacks around 1900, but the 1916 map shows that most of the large fires occurred near the railroad lines to the north. There were smaller fires in Perkins Clearing to the north (1909) and around Echo Lake (1910) in the Sacandaga Lake watershed.[35] From 1890 to 1914, various plots at Fish Mountain, between Sacandaga and Lake Pleasant, and to the north and east of Sacandaga were logged, mostly for softwoods, so that approximately half of the Sacandaga Lake watershed was logged at this time. In subsequent years, International Paper has logged portions of the northern watershed of Sacandaga Lake for pulpwood.[36] We do not have cores from the lake sediment of Sacandaga Lake to determine the effect of these activities on the lake, but the resultant increased runoff and nutrient input must have had at least some impact.

We have no firm understanding of how the large hotels in the early 1900's dealt with their waste. Early on there must have been outhouses, but by 1906 the Osborne Inn bragged of "baths and toilets...with the latest sanitary plumbing." Other hotels could no doubt make similar claims. We have no idea where the "sanitary plumbing" deposited its charge, but the lakes would have made a handy repository. After the hotel era, the summer camps surrounding the lakes had either outhouses or primitive septic systems, and the sandy soils of the area would have allowed leaching through the soil for some distance.

The New York State campground at Moffitt's Beach on Sacandaga Lake was established in late 1960's, and it currently has about 260 campsites. Its wastewater system was completed in 1968, and it treats the wastewater with a sand trickle filter, chlorinates it, and then pumps it into the lake. This process removes some bacterial food, but it doesn't extract many nutrients and the remainder eventually ends up in the lake. Municipal sewage treatment is a new luxury in the Village of Speculator since about 2000, but it currently serves only a small portion of the camps on the shore of Lake Pleasant, and none on Sacandaga.

In summary, the last century with its many large hotels and more recent lakeshore development has almost certainly contributed to the nutrient budgets of these lakes, though it is impossible to quantify its impact. We do know that the bottom layer of water in both Lake Pleasant and Sacandaga Lake has very low dissolved oxygen levels in the late summer (see Chapter 4), an indication of some degree of nutrient enhancement, whereas the nearby undeveloped and isolated Fawn Lake has adequate oxygen levels for fish throughout its entire volume all year. The low oxygen levels in Sacandaga and Lake Pleasant make these lakes inhospitable to sustainable populations of lake trout and certainly contribute to their absence in these lakes.

So Sacandaga Lake, as with all Adirondack lakes, is the result of historical factors both natural and human. Its shape and watershed chemistry were determined hundreds of millions of years ago as a result of continental drift and other geological processes. Its adjacent forests have altered its water chemistry since the retreat of the glacier 10,000-plus years ago, and they are still having an impact. Humans have been relative late-comers to the area, but they have affected the lake by altering its surrounding vegetation, exploiting its natural resources, and no doubt enhancing its nutrient loading. Fortunately, the lake ecosystem has been fairly resilient so far, but we must try to better understand these lakes so we can minimize our impact in the future.

3

Why Water is Weird

Have you ever wondered why ice floats, or how a water strider can "walk" along the surface of water and not sink? Why certain fish species must be fished deep, whereas others can be found in shallow water? Why a springtime swim in a lake can be warm on the surface but cold on your toes as you tread water? Why the space agency is so focused on finding water on extraterrestrial bodies? Water is critical to life as we know it—we are made up of roughly two-thirds water, and our most metabolically active cells may be more than 90% water. Although water is such a common and important molecule in our environment, it has many strange chemical characteristics that set it apart from other molecules of similar size.

When you add heat to a liquid, its molecules, which are always in motion, speed up. If you add enough heat, eventually some molecules will be moving fast enough to free themselves from their liquid neighbors to form a gas. The temperature at which a liquid changes phase into a gas is known as its boiling point. As a rule of thumb, the boiling point of a compound increases as its molecular size, or mass, increases. This makes sense since the more massive the molecule, the more heat energy will be needed to speed up the molecular motion so that these larger molecules can free themselves from chemical attractions that keep them in the more dense liquid form. For example, methane (CH_4) has a molecular mass of 16 and a boiling point of minus 260 °F, ethane (C_2H_6) with a mass of 30 boils at minus 126 °F, and hydrogen sulfide (H_2S) with a mass of 34 boils at minus 80 °F. Therefore we might assume that water, with a molecular mass of 18, would boil at a temperature somewhere between methane and ethane, but much closer to methane. Instead, water boils at +212 °F—over 400 °F higher than we would have expected considering molecular mass alone. Why should it take so much more energy to boil water than methane? Obviously something else is going on here besides the effect of molecular mass, and this is what gives water its distinctive characteristics.

The Chemistry of Water

To explain this apparent anomaly, we must take a brief foray into chemistry; but bear with me here, it won't take long. The answer to this paradox is that there are weak *inter*molecular bonds (called **hydrogen bonds**) between water molecules that tend to hold them together, and these bonds are absent from most other molecules like methane. It is this ability of water to form these intermolecular hydrogen bonds that gives water its distinctive characteristics. These bonds give water a very large surface tension, so that organisms such as water striders and water beetles can actually support themselves above the water surface. In fact, if you grease up a metal sewing needle, it will also float on the water's surface, held up by the strong surface tension.

Water also has a relatively high viscosity compared with other liquids that are made up of small molecules. This allows microscopic organisms in a lake to remain suspended longer in the water column even though gravity causes them to sink. These bonds also explain why it takes so much more energy in the form of heat to vaporize water, and hence its relatively high boiling point compared with other molecules its size.

However, the effect of hydrogen bonding in water that is most relevant to lake ecology is its influence on density. Most materials—gasses, liquids and solids—expand when they are heated and contract (become more dense) when they are cooled. However, if you measure the density of water relative to temperature, an anomaly occurs. As water is cooled and approaches its freezing point (0 °C), it actually becomes less dense (Table 3.1). (For a refresher on the different temperature scales and the metric

The Secret Life of a Lake

Table 3.1
The density of water relative to temperature.

Temperature	Density (g/cm³)
30 °C (86 °F)	0.99567
20 °C (68 °F)	0.99823
10 °C (50 °F)	0.99972
4 °C (39 °F)	1.00000
0 °C (32 °F)	0.99986
Ice	0.91680

system, see Appendix I.)

Note that water is at its most dense at 4 °C (39 °F), and its density decreases as it approaches 0 °C (32 °F). Ice is the least dense of all, and of course, this is why ice floats on water. But how can this happen? If you have been reading carefully, you must know that hydrogen bonds are involved. (For a detailed discussion of the unique chemistry of water, please see Appendix III.)

The Water in Lakes

Now let's apply this concept to what actually occurs in a lake. Lakes in temperate regions go through a typical annual freeze–thaw cycle. As the sun's intensity decreases in the fall, less energy is available to heat the lake. Evaporation and other avenues of energy loss are always occurring at the lake surface, cooling it; so without a compensatory heating by the sun, the lake water cools. As it cools, colder denser water will sink to the bottom, until the entire lake reaches 4 °C (39 °F). Further cooling at the surface will produce less dense water, which will remain at the surface. Eventually the surface will cool enough to freeze, and the less-dense ice will float on the lake's surface. Most of the water below the ice all the way to the bottom of a deep lake will remain at 4 °C (39 °F) for the rest of the winter, except for the thin layer just adjacent to the ice that is colder.

Think of what would happen if water acted like all other liquids in which the solid form is denser than its liquid form. As the lake cooled, the colder, denser water would sink to the bottom. Ice would form on the bottom of the lake, and in colder regions, the entire lake would eventually freeze, killing its inhabitants. In the summer, only the top of the lake would thaw, because the warmer, less dense water would remain at the top. In regions such as the Adirondacks, the bottom of the lake would never thaw. So we can thank water's temperature–density anomaly for allowing our lakes to retain their biological diversity throughout the entire year.

Winter Ice

We'll now return to the actual situation where the cooled lake surface is forming ice. For ice to form, the water molecules must develop an intricate lattice structure (see Appendix III for a detailed discussion). If the water is in motion, such as in a stream or a windy lake, ice cannot form and the water may even become supercooled to *below* the freezing point. If windy conditions persist, small clumps of ice crystals may form and collect on the downwind shore. However, in most cases a cold, windless night will allow the water to freeze, first in protected bays with less water movement, and last in areas where water movement persists, such as at the mouths of tributaries or outlets. If the water is supercooled, the entire lake can freeze over in a single windless night. Once an ice sheet forms, the lake water is isolated from wind action.

As an ice crystal forms, it reflects the angles of the bonds between the water molecules, and the result is a hexagonal shape. This is the reason that snowflakes look like elaborate hexagons, and ice crystals form in lakes as hexagons. In a lake, ice crystals are larger if the water was very still at the time of formation and smaller if it was in motion. A typical ice crystal dimension in a lake is about an inch (2.5 cm) across. Fresh ice has few gas bubbles in it and is very transparent. Since the water beneath the ice is usually dark, this ice is known as **black ice**, and it is the strongest of all the forms of ice. It conducts the cold well so that the crystal continues to grow on its bottom, forming long, hexagonal columns.

The Strength of Ice

Whenever I encounter a frozen lake (or even a frozen puddle), I am always tempted to test the strength of the ice. The U.S. Army Cold Regions Research Lab in Hanover, New Hampshire, has published the data in Table 3.2, which show the relationship between ice thickness and its ability to support weight.[1] The values listed in the table are for black ice.

Eighteen inches of ice is not uncommon on Adirondack lakes and can easily support a large truck. The problem is, however, that we can't tell how

Table 3.2
Minimum black ice thickness to support a load.

Load (tons)	Required Thickness (inches)
0.1 (200 lbs)	2
1.0	4
5.0	9
10.0	13
20.0	18

Note. One ton equals 2,000 lbs.

thick the ice is at any given point on a lake by just looking at the surface. As previously stated, water movement at inlets and outlets will make the ice thinner, and any underwater current will bring up warmer 4 °C (39 °F) water to further thin the ice. Underwater springs of even small flow rates will also thin the ice above them. Every winter, localized thin ice on Adirondack lakes claims snowmobiles, trucks and lives.

Black ice will absorb the sun's radiation and will warm up on sunny days. When this happens, the ice will expand, like all solids, and it will crack and buckle, forming expansion ridges. Almost everyone who has been on a frozen lake during a sunny winter day has experienced the unnerving sound of cracking ice. These expansion ridges can lift the ice high enough to send a speeding snowmobile airborne, sometimes with fatal consequences. When the temperature cools down at night, the ice will contract and may even form areas of open water along the cracks. This daily expansion and contraction of the ice sheet gathers rocks and boulders in the shallow shoreline areas and eventually pushes them toward the shore, helping to produce the rocky shorelines found on most Adirondack lakes. This thermal ice movement will also crush docks or other structures at the water's edge.

The Impact of Snow

Eventually snow will fall on the ice, creating a highly reflective layer. Solar penetration into the water column will be cut to levels that will no longer be able to support photosynthesis. Although the water temperature below the ice is 4 °C (39 °F) or less, life continues, as any ice fisherman will tell you. Metabolic rates are low, but not zero. Not only are the fish active, but microscopic animals, bacteria and fungi continue to function. All these creatures consume oxygen from the water. During the warmer ice-free months, oxygen in the water column is replaced by diffusion from the atmosphere and photosynthesis by algae and rooted plants. However, after ice forms, the water below the ice is isolated from the atmosphere, and snow-covered ice will reduce oxygen-producing photosynthesis to negligible levels. Because all the organisms below the ice are consuming oxygen through respiration, the oxygen content of the water will decline. Depending on the volume of water in the lake and its depth, the length of time it is covered with ice, and the nutrient levels in the lake, oxygen concentrations may decrease to levels lower than fish can tolerate, and fish kills may result. Winter fish kills are much more common in shallow, nutrient rich lakes, and to my knowledge none has ever occurred in Sacandaga Lake.

Another effect of snow load on the ice is that the weight of the snow will depress the ice, pushing the ice surface below the lake water level. The resulting water pressure will push liquid water up though cracks and fissures in the ice so that there will be a layer of liquid water between the hard ice below and the frozen snow on top.[2] If the weather is cold enough, this snow slush will freeze, making a layer of frozen slush. As the weather warms, sun penetrating the snow may thaw this slush layer, producing the anomaly of liquid water just under a thin crust of frozen snow and slush. Walking on the lake at this time, you may penetrate the top layer and get a wet foot, but, if you're lucky, the ice layer below will still be intact and you'll only be mildly inconvenienced instead of at risk of drowning.

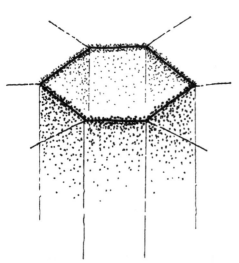

Figure 3.1
A single ice crystal in an ice sheet. Note the elongated hexagonal shape and the concentration of dissolved ions (dots) at the interface between crystals.

When the ice crystals form, they start in the center of the hexagon and grow outward toward the periphery.

The Secret Life of a Lake

Crystal formation requires pure water, so as the crystal forms, it excludes "impurities" from the water, forcing them to the outside. The result is that the region of contact between adjacent crystals has a higher concentration of ions such as calcium, chloride, and nitrates than the central part of the crystal (Figure 3.1).

(An aside: This principle is used in the production of applejack, an alcoholic beverage made from cider. Cider is allowed to ferment, then poured into a container that is left out to freeze in the winter. The cider freezes starting at the surface of the container, and it excludes the "impurities," which concentrate in the center. In the case of fermented cider, a major impurity is alcohol, which is concentrated many times. The center of the container is then tapped, and the applejack extracted.)

Spring's Arrival

Eventually the sun's radiation intensity increases as spring advances, and the ice begins to melt. Usually the land around the shoreline becomes free of snow before the lake thaws, and it absorbs more of the sun's energy than the snow-covered lake. This results in a warming of the shoreline, and often an open-water moat forms around the lake's edge. As the ice sheet warms, it does not melt evenly. The first ice to melt is that with the lowest melting point. Remember the layer between the columnar ice crystals with the higher concentration of ions (Figure 3.1)? When you add salt or other soluble compounds to water, it decreases the water's freezing point. (Antifreeze in your car radiator does the same.) Therefore as the ice warms, the thin layer between the crystals will melt before the main body of the crystal. When this occurs, the ice loses its transparency and becomes clouded. Ice at this stage is often called gray, or "rotten ice." Once this happens, the ice loses its strength, since now the vertical, parallel crystals are no longer held together and can slide against each other.[3] A rotten ice sheet a foot thick cannot support your weight, whereas 13 inches of black ice can support 10 tons (Table 3.2).

Wave action along the edge of a rotten ice sheet will break the elongated crystals away, and since these long, thin crystals look like candles, we say that the ice has "***candled***." Sometimes this breakup is accompanied by a tinkling sound as the candles clink against each other—one friend described this sound as a faint wind chime. A large candled ice sheet can break up very rapidly, so fast that some people think that the ice actually sinks in the spring. Of course ice can't sink, but it breaks up into the long, thin hexagonal crystals that are now floating horizontally in the water and are not obvious to the shoreline viewer. Thoreau described these last stages of ice melt on Walden Pond. He wrote that the ice became...

> ...honeycombed and saturated with water, so that you could put your foot through it when it was six inches thick; but by the next day evening, perhaps, after a warm rain followed by a fog, it would have wholly disappeared.[4]

Seasonal Cycles

After the ice melts, the sun's radiation will warm the surface of the lake. The narrow, colder layer closest to the ice will warm to 4 °C (39 °F), so that the entire water column will be the same temperature, or *isothermal*. Under isothermal conditions there are no density differences in the water throughout the depth of the lake, and wind action can mix the entire volume of the lake. When this occurs, it is called a *turnover*, and since it occurs in the spring, it is called the spring turnover. This period of turnover can be extended in lakes exposed to lots of wind, or it can be brief in lakes protected from the wind. As the lake turns over, its entire volume is warmed uniformly.

Eventually there will be a period of high solar input and low wind activity, and the top of the lake will warm up above the temperature of the lower layers. Since warmer water (above 4 °C) is less dense than cooler water (see Table 3.1), it will remain at the top of the lake. After this warmer water layer is formed, subsequent wind action will mix the water within this upper layer, but in most cases will not be strong enough to mix this warmer, less dense, layer with the colder water beneath. When this division of the lake into a warmer, upper layer and a cooler, more dense lower layer occurs, the lake is said to be *stratified*. The upper layer is called the *epilimnion* (*epi* ≈ above), the bottom layer is called the *hypolimnion* (*hypo* ≈ under), and the area where there is a steep gradient in temperature (a *thermocline*) is called the *metalimnion* (*meta* ≈ middle). This temperature profile is depicted in Figure 3.2. The thermocline is usually defined as the region where the temperature change is greater than 1 °C per meter change in depth (0.6 °F per foot).[5]

The formation of the thermocline is of critical importance to a lake,

The Secret Life of a Lake

because the temperature and resulting density gradient in the metalimnion produces a resistance to mixing that isolates the hypolimnion from the epilimnion and the atmosphere. In the spring, the thermocline may be fairly shallow. You may have even sensed a thermocline as you swam in a lake during a sunny spring day. When you swam horizontally along the surface, the water felt warm, but when you stopped to tread water in a vertical position, your kicking feet stirred up much colder water. As the summer progresses, wind mixing in the epilimnion drives the thermocline progressively deeper and makes it very steep (Figure 3.3, June through September). In the late summer the thermocline in Sacandaga Lake is about 20 ft (6 m) deep, and temperature differences greater than 4 °C in one meter's depth change (2 °F

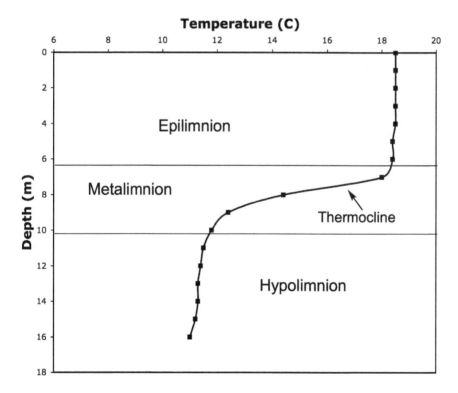

Figure 3.2
A temperature profile relative to depth in Sacandaga Lake in early September. Note the isothermal conditions in the epilimnion, the strong temperature gradient of the thermocline in the metalimnion, and the cold temperatures of the hypolimnion, even in the early fall. The surface of the lake is at the top of the graph at zero depth.

per foot) can be found in the thermocline.

To illustrate the effectiveness of the thermocline in preventing mixing, we only have to look at temperature profiles in a lake. When the lake is isothermal in the early spring just after the ice melts, the entire lake starts out at 4 °C (39 °F). If the spring is very windy, this mixing period can go on for a matter of weeks, during which the entire lake is being warmed by the sun. The length of this spring mixing period depends not only on the weather conditions, but also the shape of the lake and its orientation relative to the direction of the prevailing winds. Once the thermocline forms, however, the hypolimnion is isolated from the warming of the upper layers, and it can remain very cold throughout the summer. For instance, Sacandaga Lake is roundish in shape with a dimension of about 2 miles parallel to the direction of the prevailing winds. Nearby Fawn Lake (Figure 1.1) is as deep as Sacandaga but is smaller and narrower, and it is more protected from the

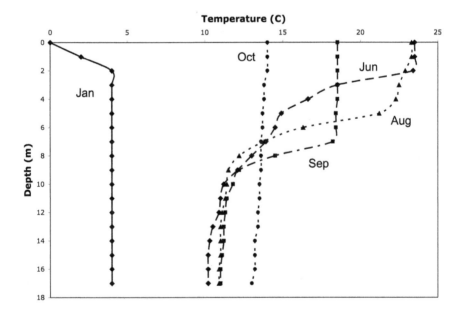

Figure 3.3
The seasonal change in the Sacandaga Lake temperature profile. Note the inverted profile in the winter under the ice (Jan), the shallow thermocline in the early summer (Jun) that deepens as the summer progresses (Aug, Sep), and the isothermal conditions during the fall turnover (Oct), which eventually cools the entire lake volume to the winter condition (Jan).

wind by adjacent hills. Fawn Lake forms its thermocline more rapidly than Sacandaga, so that the temperature of its hypolimnion has a maximum of about 6 °C (43 °F) during the summer, whereas Sacandaga's warms up to about 11 °C (52 °F) before the thermocline forms, and it stays about the same temperature all summer (Figure 3.3, June through September).

As the sun's angle decreases through the fall, the cooling effect of evaporation and long wave radiative cooling overtakes the warming of the lake by the sun, and the epilimnion cools down. Eventually the epilimnion cools to about the same temperature as the hypolimnion, the thermocline is eliminated, and the entire lake volume can once again be mixed by the wind and convection (See Figure 3.3, October). This is called the *fall turnover* and occurs in Sacandaga Lake at about 13 °C (56 °F) and about 7 °C (45 °F), in Fawn Lake. As cooling continues to outpace solar warming, the temperature of the entire lake decreases to 4 °C (39 °F), after which the inverse stratification mentioned previously (colder water on top of warmer) allows ice to form, and the cycle is complete.

I have limited my discussion to temperature regimes in this chapter. However the thermocline has even more important implications for the nutrient and gas (especially oxygen) distribution in a lake, which will be considered in the next chapter.

Internal Mixing

Although the major source of mixing in a lake is the wind, another, more subtle, process may be going on at times. One summer day, I had hiked into Fawn Lake and sat on the bridge over the stream at the northern end of the lake. Having not recently looked at the topographic map, I didn't know whether this stream was an inlet or an outlet to the lake. After noting the direction of the flow as indicated by floating material and the bending of the weeds beneath the surface, I concluded that it was an outlet. I continued my hike a little way up the trail, then returned to the bridge. Now the stream was an inlet! How could this be?? I stayed on the bridge a little longer, and the flow of the stream eventually reversed itself!

What I was observing was the result of a *seiche* (pronounced "saysh"), an indirect effect of the wind. If the wind blows for a considerable length of time, it piles water up on the downwind side of a lake. The amount of this buildup depends on the strength and duration of the wind, as well as the shape, orientation and depth of the lake. When the wind stops, the water sloshes back to the other end of the lake, just as water scooped to one end of a bathtub sloshes back to the other end after the scoop leaves the water (Figure 3.4). The water sloshes back and forth until it finally comes to rest. Surface seiches on small lakes are very hard to measure. Their amplitude is small, and this small change in water level is impossible to separate from wave action, which is of much greater magnitude. On a lake the size of Sacandaga, its amplitude would be only a few millimeters (0.2 in), and its oscillation period about 15 minutes. However, on Lake Erie, a long, shallow lake with its axis approximately parallel to the prevailing winds, a surface seiche with an amplitude of about 8 ft and a period of 13 hours has been observed.[6]

Another component in the equation describing seiche dynamics is the difference in density of the two interacting media—in the case of the surface seiche, air and water. However the pile up of water on the downwind end of a lake depresses the thermocline, so when the wind subsides, the thermocline also oscillates. But in this case the difference in the densities of the two interacting media—warmer epilimnion water and cooler hypolimnion water—is much less than the air–water interface at the surface, and the resulting internal seiche has a much larger amplitude and longer period of oscillation (Figure 3.4). For instance in Lake Mendota in Wisconsin with a length of 5.4 miles, its surface seiche is a few millimeters with a period of 26 minutes, but its internal seiche can be about 5 ft with a period of 10 hours.[7] The turbulence between the epilimnion water and the hypolimnion water with internal seiches can transport some materials between the two layers.

I have not measured internal seiches in Sacandaga Lake—they are transient and require much instrumentation. But I have seen a video of their effect on the sediment in Lake Champlain. This long (108 miles), narrow (11 miles at widest point) and deep (400 ft) lake has internal seiches with amplitudes as high as 132 ft and periods of 4 days.[8] Where the sediment is at the same depth as the thermocline, currents as fast as 16 inches per second disturb and displace the sediments, which can release nutrients into the water column and disrupt the community of bottom-dwelling animals.

Light

The major source of heating for a lake is the sun. The various wavelengths of solar radiation are absorbed by water differentially. For instance, one meter of water absorbs more than 90% of the infrared radiation, but only about 2% of the blue light (Table 3.3).[9] If you have ever taken color photographs under water, you will notice a blueish tint to your pictures that increases as you dive deeper. Red light does not penetrate as deep as the blue light.

Light intensity varies with depth in a homogeneous water column in a negative logarithmic relationship. That means that for every meter's depth change, the same proportion (for example, 50%) of the light is absorbed (Figure 3.5). A semi-log plot of light intensity versus depth should generate a straight line. If it doesn't, it means that there is a layer in the water column with higher light absorption, such as a layer of algae or debris sitting on the sharp density gradient in the thermocline.

Measuring underwater light intensities does have its problems. Underwater pho-

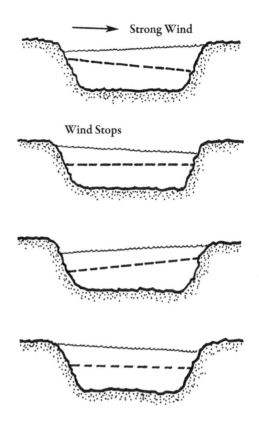

Figure 3.4
A diagram of surface and internal (dashed line) seiches of a hypothetical lake. Note the larger amplitude and slower period of the internal seiche of the hypolimnion relative to the surface seiche. The amplitude of the surface seiche is exaggerated in this diagram.

tometers are expensive and somewhat fragile under field conditions. However, a less expensive and simpler method can be used to estimate water clarity, called a ***Secchi disc***, developed a century ago by Italian physicist Angelo Secchi.[10] Secchi discs are weighted discs painted with alternate black/white quadrants; they are usually 20 cm (8 in) diameter (Figure 3.6). They are lowered into the water on a calibrated line, and the observer notes the depth at which the disc disappears. The disc is lowered farther and then

Table 3.3
Light absorption by a meter (3 ft) of distilled water as a function of color.

Color	Percent Absorption
Infrared	90
Red	42
Orange	30
Yellow	8
Green	3
Blue	2

raised until it reappears. The average of these two depths is called the Secchi depth. This crude technique is surprisingly repeatable and has been used extensively world-wide. On average, the depth coincides with about the 10% illumination level. Secchi depths range from 132 ft in Crater Lake in Oregon, one of our clearest lakes, to a few inches in heavily polluted lakes.[11] I have been on a lake in the Hudson-Mohawk Valley with a Secchi depth of only 4 in! In Sacandaga Lake the summer Secchi depths range in the 13 to 20 ft (4 to 6 m) level, with a rare maximum of 23 ft (7 m).

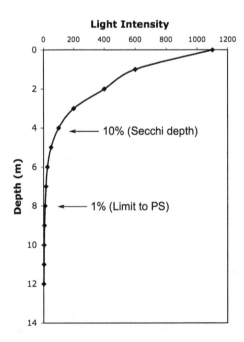

Water transparency is an important lake characteristic for both biological and aesthetic reasons. Subjectively, most people prefer a lake with clear water to one with murky water. Biologically, water clarity is

Figure 3.5
The light profile relative to depth in July in Sacandaga Lake. Note the steep decline in the upper region of the lake, the approximate level of Secchi depth (10%), and the approximate lower boundary for photosynthesis (PS) at 1% depth.

The Secret Life of a Lake

Figure 3.6
A Secchi disc used to measure water transparency.

a reliable indicator of the nutrient status and photosynthetic productivity of the lake.

Water clarity is affected by anything suspended in the water, as well as colored dissolved materials. Clay particles in southeastern lakes decrease clarity, as do dissolved organic humic acids from marshes that give northern lakes fed by tributaries passing through wetlands a slight tea color. An example of the latter in Sacandaga Lake occurred in July of 2006. In early July, about 6 in of rain fell in a day and a half. The lake level rose about two feet in a few days, and the tributaries' wetlands were washed free of all their deeply colored waters that were then deposited in the lake. Figure 3.7 shows the Secchi depth over the seasons of 2005–7. Note that the water transparency decreased significantly in July 2006, relative to other years, due to the influx of this marsh water, and it didn't recover until late in the season. Also note that the clarity in 2006 was actually better than the other years before the rain event.

However the most important material in the water affecting transparency, especially in the Adirondacks, is the particulate algae population in

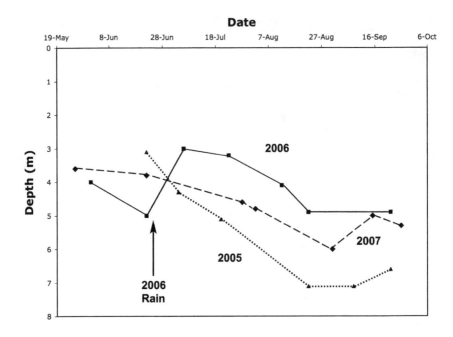

Figure 3.7
Water transparency during the ice-free seasons of 2005–2007 in Sacandaga Lake. Note that the lake surface is at the top of the graph (at 0 m), and the greater the Secchi depth, the greater the water transparency. The decrease in transparency as shown by Secchi depth in early July 2006 was the result of a heavy rainstorm and the resultant influx of tea-colored humic substances from tributary wetlands.

the lake. Because of this, water transparency is often correlated with algal productivity, which reflects the nutrient status of the lake. Therefore observations of water transparency are critical to understanding the biological dynamics in a lake, a fact discussed in more detail in later chapters.

So the next time you look out over your favorite lake's peaceful surface, consider what might be going on underneath. There are really two lakes here: the warmer epilimnion and the cooler hypolimnion, with very different temperatures, nutrient levels and biological communities. The hypolimnion may even be sloshing back and forth in an internal seiche. There is also most likely a group of organisms that have adjusted their densities so they can stay at the depth of the thermocline and not have to fight gravity and the tendency to sink below the light level needed to survive. In other words, more interesting things are going on than meet the eye.

4

CHEMICALS IN THE WATER, BOTH GOOD AND BAD

IN THE EARLY DAYS OF organic farming, vendors would advertise their products as "Free of chemicals." Of course that would be impossible, since the food itself is made up of chemicals, as are we humans. Yet many still think of "chemicals" as being bad. In reality, it is actually difficult to tell if a given chemical is beneficial or detrimental to life, since that depends on the level of exposure. Even water, if taken in a large enough volume in a short enough time, can be lethal to humans. Heavy metals like copper and zinc are toxic at moderate levels, yet are required at very low levels for the proper functioning of certain enzymes. Some chemicals can be benign when they enter an ecosystem, but be transformed within that system into a toxic form. In this chapter we'll look at many of the chemicals that can affect an aquatic ecosystem—their origins, transformations and effects on resident organisms.

Knowing the concentrations of these chemicals is critical to understanding their roles within the system. The most frequently used concentration unit is "parts per million" (ppm). This describes one unit of substance dissolved in one million units of solvent. In the metric system, this translates to one milligram (mg) in one liter (1,000 g) of water, or 1 mg/L. To make this value more concrete, visualize an Olympic-sized swimming pool 50 m long, 25 m wide and 2 m deep (165 × 83 × 6 ft). Such a pool will have a 2,500 m³ or 2,500,000 L volume. A liter of water has a mass of one kilogram (see Appendix I), so we would need 2.5 kg of something, let's say sugar, in that 2.5 million kg of water in the pool to make 1 ppm. The 2.5 kg of sugar is a little over 5 lbs, so one five pound bag of sugar mixed evenly in that large swimming pool would give an average sugar concentration of about one ppm. One part per billion (ppb) would be one thousand times less sugar (2.5 g, or a mass equal to about six paper clips) in the same pool. Modern analytical techniques can routinely measure compounds at these low levels.

Gasses

Oxygen

Atmospheric gasses like oxygen and nitrogen readily dissolve in water, so all of them can be found in lake water. Cold water holds more dissolved gasses than warm water, one reason some fish species with high requirements for oxygen (like trout) prefer cooler water. Water can even become supersaturated with gasses, but this is an unstable condition and the gas will bubble out of solution if disturbed.

Gaseous diatomic nitrogen (N_2), the most common atmospheric gas (at 78%), is metabolically inert and won't be mentioned further, but atmospheric oxygen (21%) is a major source of oxygen in a lake. However, it is not the only source. Photosynthetic organisms can use the radiant energy from the sun to create chemical energy (think food), which is used by all organisms in the lake for their own metabolism. Organisms that can carry out photosynthesis are called ***autotrophs*** (*auto* ≈ self; *troph* ≈ to feed) because they can "feed" themselves. Organisms that cannot photosynthesize are called ***heterotrophs*** (*hetero* ≈ different or other), because they must ingest or absorb high-energy compounds (food) from their environment. An approximate formula for the autotrophic photosynthetic reactions takes carbon dioxide and water and produces oxygen and glucose (a sugar) using the radiant energy from the sun. Photosynthesis occurs as follows:

Carbon Dioxide + Water + Light Energy →
Oxygen + Sugar (Glucose)

Chlorophyll is required for this transformation, and algae and vascular plants are by far the most common organisms in the lake that can carry out this reaction. (In some lakes photosynthetic bacteria can be present.) Photosynthesis, therefore, can increase the oxygen levels in a lake, and it will produce the high-energy sugars required to "feed" all the organisms in the aquatic ecosystem.

While higher plants and algae are the main organisms that can photosynthesize, all organisms, including the algae and plants, must carry out cellular respiration. The respiration reactions are the exact opposite of the photosynthetic reactions in that they use the sugars and break them down to produce high-energy intermediate compounds that an organism needs for its own metabolism—to grow, move, make proteins, reproduce, and so on—in other words, to live. These reactions *use* oxygen to break down the sugars, thereby reducing the dissolved oxygen levels in the water. Cellular respiration occurs as follows:

$$\text{Sugar (Glucose)} + \text{Oxygen} \rightarrow \text{Carbon Dioxide} + \text{Water} + \text{Metabolic Energy}$$

To illustrate how much energy is stored in the bonds of the sugar molecule, the above reaction is essentially the same as the reaction of burning firewood in your wood stove. The sugar molecule is glucose, the same molecule that makes up the cellulose of wood. Once you start your fire, the glucose in the wood is combined with oxygen, producing carbon dioxide and water, and the energy released from the glucose bonds is released in the form of heat, which then warms your home. In the respiratory pathways of an organism, much of the energy is transferred into other high-energy molecules in an orderly series of enzyme-mediated reactions, which enable the organism to build complex molecules, power muscles, and grow.

Seasonal and Stratification Variations in Dissolved Oxygen

Therefore photosynthesis produces oxygen and respiration uses it. The reverse occurs with carbon dioxide. How do these reactions affect a lake? In the spring after ice melt, the spring turnover mixes the lake throughout its volume. The water is cold, thereby dissolving a maximum amount of oxygen from the atmosphere, and the turnover mixes the oxygen evenly throughout the water column. After the thermocline forms, the epilimnion is isolated from the hypolimnion, and the atmosphere can no longer supply oxygen to the hypolimnion. Photosynthesis can take place, but the light levels in the hypolimnion are usually too low to allow much photosynthesis (Figure 3.5).

There are many heterotrophic organisms in the hypolimnion that continue to respire—bacteria, fish, microscopic animals, and fungi—and the sediment contains a very active group of organisms decomposing organic material. In addition, there is a constant "rain" of organic material in the form of settling algae, dead animals and organic debris that sinks from the epilimnion into the hypolimnion. Much of this material is decomposed by bacteria and fungi within the hypolimnion, even before it gets to the sediment.

All this metabolic activity requires oxygen, and the two main sources

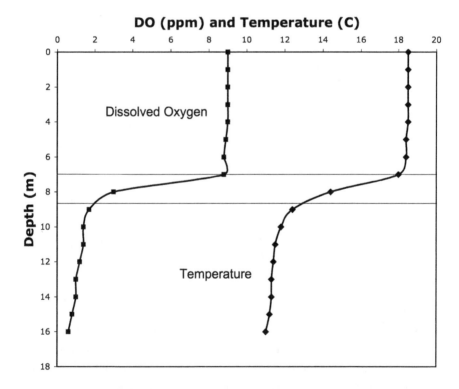

Figure 4.1
Dissolved oxygen (DO) and temperature profile in Sacandaga Lake in September. Note the effect of mixing in the epilimnion and the abrupt decrease in DO at the thermocline. At this time, the hypolimnion is very close to anoxic.

of oxygen, the atmosphere and photosynthesis, are not available. As a result, the dissolved oxygen levels in the hypolimnion decrease as the summer progresses. The rate of this decrease depends on the nutrient levels in the lake, the volume of the hypolimnion and the amount of organic material in the sediment. The hypolimnion can eventually become essentially devoid of oxygen, or *anoxic*, in many lakes. Figure 4.1 shows a profile of dissolved oxygen (DO) and temperature measured in September in Sacandaga Lake.

Note that the DO levels are high and consistent throughout the depth of the epilimnion due to continuing photosynthesis, contact with the atmosphere, and mixing by wind within the layer. However, at the thermocline the DO decreases dramatically. The graph in Figure 4.1 illustrates the powerful effect of the thermocline in isolating the hypolimnion from the epilimnion, not only thermally but also chemically.

In some nutrient-rich lakes, the hypolimnion can become anoxic within weeks of the formation of the thermocline. In Collins Lake, a small urban lake in Scotia, New York, the hypolimnion becomes anoxic within a month of ice melt—by early May—and it remains that way until the fall turnover in late October. This lake has a relatively small volume in its hypolimnion and much organic material in its sediment. An anoxic hypolimnion has definite implications for the type of fish that can survive in a lake as well as the mobilization of certain sediment chemicals, as we will discuss shortly.

Most Adirondack lakes do not have such a rapid drop in hypolimnion oxygen levels, but hypolimnetic DO will decrease in all of them. In Sacandaga Lake the hypolimnetic DO drops more slowly, but there is always a short period of very low oxygen levels in mid fall (Figure 4.2). Note in the figure that the lake has the most oxygen in the spring after the spring turnover (May) when the water is cold, about 13 °C (55 °F) at the top to 8 °C (46 °F) at the bottom. By June, some DO depletion has already occurred in the hypolimnion, and there is a DO peak at 3 m (10 ft), the position of the thermocline at this time (see Figure 3.3 for the June temperature profile). Some species of algae can adjust their density so that they will be able to remain suspended on the density gradient provided by the thermocline, and their photosynthesis has produced this DO peak.

By August, the thermocline has dropped (see Figure 3.3) and the DO has continued to decrease in the hypolimnion. Dissolved oxygen levels this low discourage most fish from remaining long in the hypolimnion. By September, the hypolimnetic DO has dropped to nearly anoxic levels, and at

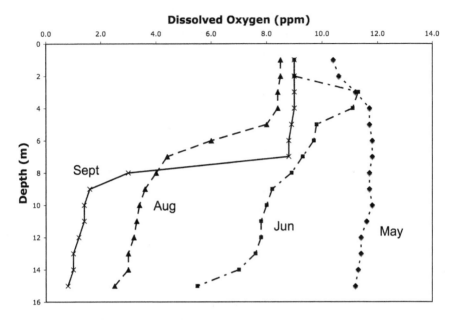

Figure 4.2
Dissolved oxygen (DO) profiles during a summer season. Note the saturated conditions during spring turnover (May) and the slow decrease in hypolimnion DO as the season progresses. In October, the fall turnover will resaturate the entire water column. Compare the positions of the DO decline in each sample date with the position of the thermocline on the same dates in Figure 3.3. The peak in DO at about 4 m in June is due to a layer of algae suspended on the thermocline.

these levels most cold-water fish will be limited to the thermocline. However, by mid-October, the thermocline has been eliminated and the lake is isothermal and turning over (Figure 3.3), so that DO levels throughout the lake will be back to saturated levels.

Therefore for at least a short time in late September and early October, Sacandaga Lake has a nearly anoxic hypolimnion. The same is true of Lake Pleasant and many other Adirondack lakes of comparable size and depth.[1] However, Fawn Lake's hypolimnion does not become anoxic, nor does nearby Piseco Lake's. The latter is a larger, deeper lake than Sacandaga, with a very large volume in its hypolimnion, and it would be surprising if it became anoxic. Fawn Lake is smaller than Sacandaga and about the same depth. Its hypolimnion/epilimnion volume ratio is somewhat larger than Sacandaga's, but the main difference is the lack of residential development

The Secret Life of a Lake 53

in its watershed.

Both Fawn Lake and Piseco Lake have a reproducing and continuing population of lake trout, a well-known cold water species, whereas Sacandaga Lake and Lake Pleasant do not. We know that both of the latter lakes had lake trout in the 19th century from historical documents (see Chapter 2), and the 1933 New York State Biological Survey of the Upper Hudson Watershed listed lake trout as "common" in both lakes.[2] It is tempting to relate the loss of this fish species to human development around these two lakes, with its possible nutrient enhancement resulting in anoxic hypolimnions. However if shoreline development were the only reason for the low oxygen levels, we would expect that Lake Pleasant, which had a longer and more intensive shoreline development pattern than Sacandaga, would have lower hypolimnetic DO levels than Sacandaga, but they both deplete their hypolimnetic DO at about the same rate during the summer. Would these low DO levels be enough to decrease the survival of the lake trout? Overfishing during the big hotel era certainly contributed, but low DO levels in the hypolimnion could have been the last straw.

Carbon Dioxide

All atmospheric gasses dissolve in the lake water, and most remain in their original chemical form, but carbon dioxide (CO_2) is an exception. When it comes into contact with water, a chemical reaction occurs joining the CO_2 with water (H_2O) to form carbonic acid (H_2CO_3). Carbonic acid then dissociates to form two hydrogen ions (H^+) that produce a mild acidity to the water, decreasing its pH to about 5.6. (See Appendix II for an explanation of pH.) Atmospheric pollutants can cause the rain to be even more acidic, discussed later in this chapter.

During the day, photosynthesis by autotrophs will extract the carbon dioxide from lake water, and the lake pH will increase (become more basic). At night, respiration from both the heterotrophs and autotrophs will produce carbon dioxide, and the pH will decrease (become more acidic). As a result there will be a diurnal cycle of pH in the epilimnion, depending on the amount of photosynthesis going on, and the cycle will be more pronounced in lakes with much higher nutrient availability than Sacandaga. The hypolimnion in most lakes will be more acidic than the epilimnion because it will contain more carbon dioxide from respiring organisms and the lack of photosynthesis due to light limitation. Lowering a pH probe into the hypolimnion of Sacandaga Lake in the summer shows that it is at least one pH unit more acidic (pH about 6) than a surface sample (pH about 7).

Nutrients

Limiting Nutrients

One of the important concepts in ecology is that of the *limiting factor*, or the variable that limits the growth of an organism or population at a given time. In the context of nutrients, it is the nutrient that is least abundant in the environment *relative to the needs of the organism*. The italicized phrase is critical since an organism doesn't need all nutrients at equal concentrations.[3] For instance, organisms require relatively high levels of nitrogen, because they contain a lot of proteins that are made up of about 15% nitrogen. They require less phosphorus and even less zinc, which is required as a cofactor for just a few enzymes. Even the micronutrient iron can limit marine productivity in some reaches of the ocean.

In freshwater systems, nitrogen and phosphorus are the main limiting nutrients, but which of the two is limiting at any given time will depend on the makeup of the algal community, the stage of their growth cycle and the chemical composition of the lake. Determining which nutrient is limiting in a lake allows us to control algal blooms by limiting the entry of that nu-

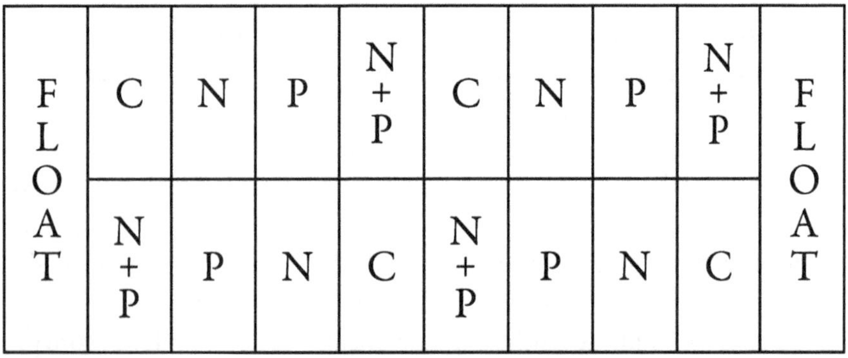

FLOAT	C	N	P	N+P	C	N	P	N+P	FLOAT
	N+P	P	N	C	N+P	P	N	C	

Figure 4.3
A top view diagram of the floating frame used in the nutrient limitation experiment. Each compartment held a 40 L clear plastic garbage bag containing lake water with the noted added nutrients (N = nitrogen, P = phosphorus, C = control).

trient into the lake ecosystem.

To determine whether nitrogen (N) or phosphorus (P) limits algal growth in Sacandaga Lake, my students and I conducted the following experiment. We anchored a floating wooden frame in the lake containing 16 rectangular openings (Figure 4.3). Into each opening we stapled a 10 gallon (40 L) clear plastic garbage bag, and we filled each bag with lake water. We left four bags alone as controls, and we put additional nitrate (N) ions into four, additional phosphorus (P) into four, and both N and P into the last four.

We left the experiment in the lake for a week. The water in the bags remained at the same temperature and light intensity as adjacent lake water, and the bags contained the same algal community as the lake at the beginning of the experiment. Our assumption was that whatever occurred during the week inside the bags would have also happened in the lake had we used the same nutrient treatments. At the end of the week, we took a sample of the water from each bag and analyzed it for chlorophyll content, which estimates the size of the algal population (Table 4.1).

There is no significant added growth of the algal community with the addition of nitrogen, whereas there is a large increase with added phosphorus. There is a small, but statistically insignificant, increase when nitrogen is added to the phosphorus treatment. Therefore phosphorus is the limiting factor to algal growth at the time of the experiment in Sacandaga Lake, as it is during most of the summer in most Adirondack lakes.

In lake water, nutrient levels do not remain constant over time. During spring turnover, the water column is recharged with nutrients and thoroughly mixed. However, during the growing season, algae will take the nutrients out of the water—algae can concentrate phosphorus from the water

Table 4.1
Average chlorophyll content in water from four nutrient treatments.

Treatment	Chlorophyll (µg/L)
Control	6.4
Added Nitrogen (N)	6.7
Added Phosphorus (P)	13.5
Added N + P	15.3

up to 10,000 times—and they will either settle out of the epilimnion or be eaten by small animals that will eventually die (or be eaten). Eventually most of the nutrients will settle out of the epilimnion and be deposited in the hypolimnion. So during the summer, low nutrient levels in the epilimnion will typically limit algal growth. An example of this seasonal nutrient depletion cycle is shown in the nitrate concentration of Sacandaga Lake (Figure 4.4).

Note that the two periods of high nutrient levels coincide with turnovers in the spring and fall. In the spring, water temperatures are cold, so algal blooms from the high nutrient levels are muted. However, in the fall water temperatures are more moderate, and blooms are more prevalent. If the water is mixed by the wind in the fall, these blooms may be unnoticed to the casual observer, but they will be obvious in reduced Secchi disc transparency depths (see Chapter 3). If the winds are calm and the blooming

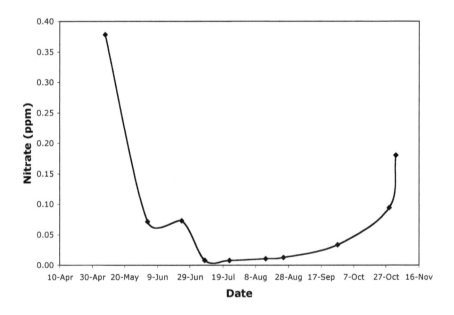

Figure 4.4
Epilimnetic nitrate levels in Sacandaga Lake over a growing season. Note the high levels during the spring turnover, the low levels in the summer, and the increasing levels at the fall turnover. Phosphorus levels would show a similar curve, but the concentrations would be lower.

The Secret Life of a Lake

algae are able to produce gas vacuoles, a green layer may form floating on the surface in portions of the lake.

Phosphorus

As shown above, phosphorus is the most common limiting nutrient in freshwater lakes. Organisms use it to make membranes, bones, and many molecules critical to life, such as nucleic acids (DNA and RNA) and energy transfer (for example, ATP) and signaling molecules. In pristine areas, phosphorus comes from the weathering of bedrock—more from sedimentary rock and less from granitic crystalline rock—and from decaying organic matter. A small amount comes from dry atmospheric deposition such as dust, but very little comes from rain.

Once in an ecosystem, phosphorus undergoes complex recycling. In the epilimnion, most of the phosphorus is found inside organisms. Since it is so rare in the form dissolved in water (phosphate, PO_4^{3-}), organisms cycle it among themselves. After the algae absorb the dissolved phosphorus, they are ingested by small filter feeding animals who use the phosphorus for their own metabolism or excrete it. The excreted phosphorus can once again be absorbed by algae. Smaller animals are eaten by larger animals, and the cycle goes on. After their death, all of these organisms will settle down into the hypolimnion, so that as the summer goes by, the hypolimnetic phosphorus levels can increase. The sediment receives a lot of this organic "rain," and the creatures down there recycle it among themselves.

To illustrate how some organisms have evolved to extract this rare and important nutrient from the environment, there is a type of algae (*Elaktothrix*) that is surrounded by a sheath of clear, gelatinous material. When one of the filter feeding animals ingests it, the alga passes through the animal's digestive tract relatively unscathed, since the digestive juices work only on the sheath rather than the algal cell. When the alga is excreted, researchers found that it had absorbed some phosphorus from the digesting material within the animal's gut, so that it was nutritionally better off after being eaten. It actually reproduced more rapidly after it was excreted than if it had not been eaten at all! In this case, we might even consider this alga a phosphorus parasite on the animal.

Dissolved phosphate that comes in contact with the sediment can combine with iron ions (Fe^{3+}) found there to form a water-insoluble molecule, ferric phosphate. This takes the phosphorus out of solution so it is not available for absorption by any organism. However, when the hypolimnion becomes anoxic, the Fe^{3+} iron is chemically changed to Fe^{2+}, which transforms the insoluble ferric phosphate into a water soluble form, ferrous phosphate. The result is that the phosphate is once again dissolved in the water column within the anoxic hypolimnion. This process is called ***internal loading***.[4]

During the fall turnover, the hypolimnetic phosphate is cycled up into the photosynthetic zone where algae can absorb it and multiply, producing a fall algal bloom. The amount of phosphate released into the water from the sediment depends on the length of time the hypolimnion is anoxic. In Upper Saranac Lake, 60 miles north of Sacandaga, late summer hypolimnion total phosphorus (TP) levels were about 6 times those found in the epilimnion. During the fall turnover when the lake is mixed, TP levels in the well illuminated surface water more than doubled, resulting in a subsequent algal bloom as mediated by an increase in chlorophyll levels.[5] So far, we have not found a similar increase in Sacandaga Lake.

Most Adirondack rocks are low in phosphorus, which is why it is a major limiting nutrient there. It is also why decreasing even small human inputs of phosphorus into our lakes is so important to controlling algal blooms. Septic systems must be tested for leakage, and lakeside lawns should be minimized, or at least there should be a strip of native vegetation between the lawn and the lake. Lawns do not hold nutrients as well as deeply rooted native plants. If present, lawns should not be fertilized, since the sandy soils typical of the Adirondacks do not bind phosphorus well. If lakeshore residents are adamant about fertilizing their lakeside lawns, they should use a fertilizer with little or no phosphorus. Fertilizers are typically labeled with three numbers (for example, 20–10–15). The first number refers to the nitrogen percent, the second phosphorus, and the third potassium. So people living along a lakeshore should use a lawn fertilizer with a zero in the second position (such as 20–0–15).

Nitrogen

Whereas soluble phosphorus is typically found in only one form—the phosphate ion—nitrogen can found be in three. Nitrate (NO_3^-) is the most stable and common, but nitrite (NO_2^-) and ammonia (NH_4^+) can also be found. When organic material is decomposed, the nitrogen in the proteins is released as ammonia. In natural waters, the ammonia is transformed by naturally occurring bacteria first to nitrite, then to nitrate. These reactions yield energy, and the transforming bacteria use this energy to grow and reproduce.

In natural waters with a high oxygen content, the presence of ammonia indicates organic decomposition nearby. If a wastewater treatment facility has an outfall into a stream, directly downstream from the outfall the ammonia levels may be high. Farther downstream, the ammonia levels will drop as the bacteria do their thing, and nitrite levels will increase. Farther still downstream, ammonia will be absent, nitrite levels will be decreasing rapidly, and nitrate levels will increase until all the soluble nitrogen will be in the form of nitrate. In most lakes, virtually all the nitrogen will be in the form of nitrates, and if it is not, some detective work is in order to find the fresh source of the decomposing organic material (like a sewage outfall). However in the chemically low-oxygen environment of an anoxic hypolimnion, the nitrate may revert back to ammonium.

Atmospheric nitrogen (N_2) is extremely stable and not available metabolically. However some organisms have evolved the ability to fix atmospheric N into ammonium, which is then transferred onto a forming protein. Some free-living and symbiotic (think of legumes) terrestrial bacteria can do it, but in aquatic systems, the cyanobacteria (aka blue-green algae) are the main nitrogen-fixers. We will speak more about this group in Chapter 6.

But there is another, more recent contributor to the nitrogen budgets of lakes—acid rain. The high compression of internal combustion engines in our automobiles transforms atmospheric nitrogen into various oxides of nitrogen (for example, NO, NO_2, N_2O_2), collectively called NOx. Once in the atmosphere, NOx is transformed into nitric acid, HNO_3, one of the major contributors to acid rain. Acid rain is a significant nitrogen contributor to both aquatic and terrestrial ecosystems, and it is very difficult to control. You and I are a major source of atmospheric nitric acid in a diffuse (non-point source) way. With our automobiles, each of us makes just a small contribution to the problem, so it is difficult to convince us to make the economic sacrifice to control our emissions. As a result, atmospheric NOx and resultant acid rain have remained high over the last few decades.

Sulfur

Although sulfur is an essential nutrient, it is not required in very high levels and is rarely limiting in natural systems. It can come from rock weathering, but the major source in the Adirondacks is atmospheric from the smoke coming out of power plants. The sulfur dioxide (SO_2) in the smoke of power plants burning coal or high sulfur oils is transformed in the atmosphere into sulfuric acid, which rains out as acid rain. Since power companies are the main source of sulfur in the atmosphere, and there are relatively few of them, they are considered a *point source*, and we can pass laws to reduce their output. The Clean Air Act and its amendments in the 1970's and 1990's encouraged power plants to burn more low-sulfur fuels and capture the SO_2 before it leaves the smokestack. As a result, sulfur concentrations in the atmosphere have declined since 1980, whereas atmospheric NOx levels have remained relatively constant during the same period.[6] Further discussion of acid rain and its effects on lake communities can be found in Chapter 11.

Although most Adirondack rocks do not contain much sulfur, sulfate is the main negatively charged nutrient ion in Adirondack streams, and its source is atmospheric. In lakes with very low oxygen levels the in their hypolimnion and in marshy sediments, sulfate can be changed to hydrogen sulfide, which has the distinctive odor of rotten eggs.

Other Chemicals

Calcium

Calcium (Ca) levels in lakes are important for two reasons. The first is that calcium indicates the presence of calcium carbonate ($CaCO_3$) in the *watershed*. The weathering of either limestone, or its metamorphosed form marble, will dissolve the $CaCO_3$, and the resulting carbonate and bicarbonate ions are very good at buffering acids. With the constant drenching of the Adirondacks with acid rain, $CaCO_3$ in the watershed helps protect lakes from acidification. Fortunately for Sacandaga Lake, there is some marble in the bedrock of its watershed, so its surface pH has remained near neutral (at about 7).

The second reason relates to the ecological limits of some species which require calcium. For instance, zebra mussels require calcium for their shells, and the accepted lower limit is about 15 to 20 ppm .[7,8] Sacandaga Lake's calcium level is in the 4 to 6 ppm range, so unless the mussels evolve a lower Ca tolerance limit (which may be possible), we will be spared this molluskan pest. There are also hints in the literature that Ca levels may determine

The Secret Life of a Lake

which invasive plants can establish themselves in a lake, but the verdict is still out on this.

We usually think of rock weathering as a slow process, so to test that assumption, my students and I looked at the calcium budget of a moderate sized stream in the Mohawk River Valley, a 25 to 35 ft (8 to 10 m) wide stream called Alplaus Creek. At the time of the measurement, it was flowing at about 90 m^3 per minute (about 40 ft^3 per second) with a Ca concentration of 25 ppm. After doing the appropriate conversions and changing calcium values to limestone, we calculated that the Alplaus watershed was dissolving about 9 tons of limestone *each day* at that flow rate! One of the major concerns regarding acid rain in the Adirondacks is that the constant influx of acid is dissolving and thoroughly removing the buffering capacity of its watersheds. Because of the Adirondack's geological history, their buffering capacity is already at a low base level due to the relative rarity of marble and limestone. If acid rain persists into the future, the Adirondacks will be less and less able to cope with this ubiquitous pollutant due to the loss of this watershed buffering capacity.

Mercury

Power plants not only produce NOx and SO$_2$, but also mercury vapor (Hg). The mean residence time in the atmosphere for Hg is about a year, whereas for SO$_2$ it is only about 10 days.[9] So Hg is well distributed throughout the globe, including the Greenland ice caps. Elemental Hg is deposited throughout the Adirondacks, but when it enters wetland and lake sediments where oxygen is depleted by the bacteria decomposing organic matter, it can be transformed by these bacteria into **methyl mercury** (H$_3$CHg$^+$).[10]

Inorganic Hg is not a major health threat to organisms because it can be easily excreted, but methyl mercury is a major threat because it is fat-soluble and remains dissolved in fatty tissue. Older fish contain more methyl mercury than younger fish. For example, a two year old perch in the Adirondacks contains about 0.5 ppm mercury, whereas a 10 year old perch contains twice that level.[11] It can also become concentrated through the food web so that it is at a much higher concentration in the higher levels of the food web, as will be explained in Chapter 5. Some Canadian loons have greater than 100 ppm Hg in some tissues.[12] A more detailed discussion of mercury and its biological effects can be found in Chapter 11.

Trophic Status

The nutrient level in a lake determines its biological productivity and the total mass of organisms (*biomass*) present. Lakes are classified on this basis so that those with high nutrient levels are called *eutrophic* (*eu* ≈ well, very; *troph* ≈ to feed), whereas those with low levels of nutrients are called *oligotrophic* (*oligo* ≈ few). Those in between are called *mesotrophic* (*meso* ≈ middle). The most frequently used parameters for determining a lake's trophic status are water transparency as measured by a Secchi disc (Chapter 3), chlorophyll *a* levels to measure algal biomass, and total phosphorus levels that are usually the limiting nutrients in most lakes (Table 4.2).[13]

Some authors have slightly different thresholds for the various trophic states, but using the criteria in Table 4.2, Sacandaga Lake would be considered oligotrophic in its phosphorus level and mesotrophic in its chlorophyll and Secchi values, and therefore it would be classified as meso-oligotrophic.

Eutrophication, or the build-up of nutrients in a lake over time, is a natural process. Organic debris such as dead leaves is brought into the lake by its tributaries and deposited by shoreline forests. In addition, nutrients washed into the lake from the watershed and biomass produced within the lake will all gradually produce a deep organic sediment that, over time, will eventually fill in the lake depression. Shallow lakes become swamps, which then become forests. This process occurs very slowly in deep lakes, more rapidly in warm shallow lakes, but it occurs in all lakes. However, humans in the watershed tend to speed up the process through erosion, increased nutrient runoff through deforestation and development, and nutrient infusion from residential and industrial wastewater. Human-induced enhance-

Table 4.2
Trophic status indicators.

Parameter	Eutrophic	Mesotrophic	Oligotrophic	Sacadaga Lake
Secchi Depth (m)	< 2	2–5	> 5	4.4
Secchi Depth (ft)	< 7	7–16	> 16	15.0
Chlorophyll (ppb)	> 8	2–8	< 2	4.7
Phosphorus (ppb)	> 20	10–20	< 10	6.0

Note. "Less than" is indicated by the "<" symbol, and "greater than" by the ">" symbol.

ment of this process is called ***cultural eutrophication.***

We have all seen evidence of cultural eutrophication, when a clear lake becomes less clear due to an increase in the algal populations. Then algal blooms appear, at first infrequently but later becoming an annual occurrence. Once a lake starts down this cycle, it is difficult to reverse. If a richly organic sediment has had time to form, internal loading of nutrients will make eutrophication even more difficult to reverse. Chapter 10 and Appendix V contain discussions of a few strategies used to reverse this process, but they are difficult and expensive. The best approach is to catch a lake's degradation at a very early stage through careful monitoring. Then the probability of a successful intervention is the highest.

5
Some Basic Concepts

Before we begin to discuss the organisms present in a lake, we have to understand the vocabulary used to describe their habitats and the functional roles they play. Organisms have both a habitat and a *niche*. The former refers to where they live, sometimes called their "address," whereas the latter refers to what they do, or their "occupation." This dichotomy is a gross simplification, but is a useful memory crutch. Ecologists are known for creating fancy terms to describe things in their realm. An old ecological joke states that ecologists can never call a spade a spade—they call it a "geotome" (*geo* ≈ earth; *tome* ≈ to cut). In this book, I will try to keep the jargon to a minimum, but there are some basic terms that are necessary to understanding a lake ecosystem.

Habitats

We will start with the habitats within a lake. The zones of a lake are described by the type of organisms that grow there and the amount of light penetrating the water. At the edge of a lake where the water is shallow, vascular plants (plants with stems containing conductive tissue) grow with submerged root systems but stems and leaves emerging above the water line. These plants are known as ***emergents*** and are found at the upper edge of the *littoral zone*. An example is the familiar cattail. Deeper water supports plants that are mostly submerged, but can have floating leaves, such as a water lily, and even deeper water contains plants that are totally submerged. The lower level at which vascular plants do not have enough light to grow is the lower margin of the littoral zone (Figure 5.1). The depth of the lower limit to the littoral zone depends on the lake's water clarity. Below the littoral zone only a few non-vascular plants can survive.

The water volume extending outside the littoral zone is called the *pelagic zone*. The pelagic zone can be further divided into an upper layer where algal photosynthesis is greater than respiration, called the *euphotic zone* (sometimes called the photic zone). The lower zone where algal respiration is greater than photosynthesis due to low light intensity (usually lower than 1% surface intensity) is called the *aphotic zone*. The level where photosynthesis equals respiration is known as the *compensation level*. In Lake Tahoe, a very deep clear lake in California with a Secchi transparency

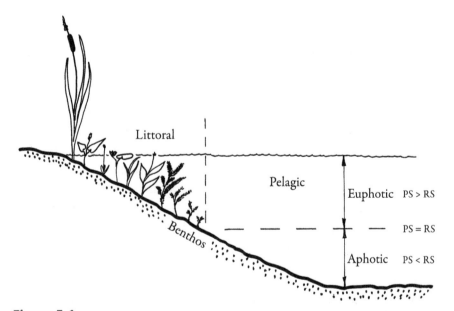

Figure 5.1
Zonation in a lake. The littoral zone is defined as the shallow area along the shore that can support vascular plant growth. The pelagic zone is the remaining, deeper volume of the lake. The euphotic zone, where algae can survive, is separated from the aphotic zone, where they cannot, by the compensation level where algal photosynthetic rate (PS) equals its respiration rate (RS). The lake bottom is called the benthos.

(see Chapter 3) of 100 to 130 ft (30 to 40 m), the euphotic zone extends to 330 ft (100 m), much deeper than the thermocline.[1] The solid bottom layer of an aquatic system, whether it is in a stream or lake, is called the *benthos* or *benthic zone*. It doesn't matter how much light is available or what the substrate is—rocks, sand or organic muck—benthos just means bottom.

Low-lying areas near a lake with standing water or saturated soils at least part of the year are called *wetlands*. These include marshes with permanent water and a typically shrubby vegetation, as well as drier areas with soils and vegetation developed by temporary soil immersion. Wetlands are valuable because they act as wildlife habitat and nutrient sinks, and they moderate flow in the associated streams. They are often protected by national, state and regional laws. Wetlands are very productive with a high photosynthetic rate, which is why they are such good nutrient sinks, and they are very rich in organic material. Decomposing organic material in wetlands produces tea-colored humic substances that can reduce water transparency in a lake (see Figure 3.7). The two major tributaries of Sacandaga Lake flow through extensive wetlands, and are deeply colored by the time they reach the lake. As a result, the lake water always has a slight tannish color.

Organisms within a lake can be described by their size and where they live. Pelagic organisms are either strong swimmers like fish, or small non-mobile or weak swimmers that are moved mostly by the lake currents. The latter are called *plankton* (*planktos* ≈ wander, drift) and can be photosynthetic algae (*phytoplankton*) or small animals (*zooplankton*). Although the plankton make up the basis for the food chain of all lakes with open water, most are microscopic. The aquatic stages of the life cycles of many insects are macroscopic—visible to the naked eye—and are known as *macroinvertebrates*. They are usually benthic.

Basic Niches

Lake organisms interact in reasonably predictable ways, and they can also be described according to their functions within a lake. Photosynthetic organisms—bacteria, algae and vascular plants—fix the radiant energy from the sun and produce the high-energy chemicals on which all the lake organisms depend. They are all lumped into the functional group called the *autotrophs*, or *producers*, since they are able to produce their own food. All other organisms are considered *heterotrophs*, or *consumers*. The consumers that eat the producers are known as *herbivores*, and those eating herbivores are the primary (1°) *carnivores*. Carnivores that eat other carnivores can be secondary (2°) or tertiary (3°) depending on their diet. The above categories refer to the organisms' nutritional roles within the ecosystem, and we call them *trophic levels* (*troph* ≈ to feed). For instance, in a lake an alga (producer) photosynthesizes, and a filter feeding zooplankton (herbivore) eats it. A free swimming insect larva (1° carnivore) eats the zooplankton, and a small minnow (2° carnivore) eats the larva. A trout (3° carnivore) eats the minnow, and an osprey (4° carnivore) eats the trout. The osprey dies, and a buzzard (a scavenger, or in this case a 5° carnivore) eats the osprey carcass.

The energy originally fixed by algal photosynthesis passes though the entire trophic structure. However, every organism requires energy to live—to build proteins, to grow, to reproduce, or in the case of animals, to move. The energy that is used for these processes is not available to the next trophic level, so fewer organisms can survive as we go up from one trophic level to the next. In addition, many organisms simply die a "natural death" and are not consumed by the next trophic level. The result is a pyramid with much more energy available in the lower levels than in the higher. A rule of thumb suggests that the energy transfer from one level to the next is about 10%. Superimposed on this pyramid is another group that feeds on dead organisms and recycles their nutrients, called *decomposers*. They include bacteria, macroinvertebrates, and fungi (Figure 5.2).

The result of this pyramid is that there are usually many more producers in any ecosystem than herbivores, more herbivores than carnivores, and very few top-level carnivores. There are rare exceptions to this generality, if one considers only the numbers of organisms or their biomass, but the relationship of energy between the trophic levels is never broken.

The trophic structure of a biological community has some important implications regarding the management of these systems. For instance, if you want to manage a terrestrial system for a top level predator like a cougar, you will need a lot of space to support only a few cougars. We will discuss in a later chapter how we can decrease one trophic level (for instance nuisance algae in a eutrophic lake) by increasing the numbers in the trophic level just above (filtering zooplankton). As far as world nutrition goes, these trophic relationships show that we could feed far more humans on earth if they were all vegetarians rather than carnivores. (However, most global human hunger is caused more by faulty distribution systems than

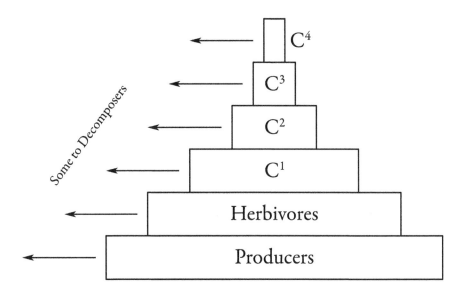

Figure 5.2
A trophic pyramid representing the producers, herbivores and various levels of carnivores. The horizontal axis represents the relative amount of energy (for instance Calories/m² day) available at each trophic level (not drawn to scale). Much of the energy in each level is used by the organisms in that level to respire, and some is diverted to the decomposers.

the total caloric content of the producers.) But perhaps the most important implication of these trophic relationships for all of us to understand is that of *bioaccumulation*, also known as *biomagnification*.

Bioaccumulation

Bioaccumulation occurs in an ecosystem when a substance is taken up by a lower trophic level and can be neither broken down (metabolized) nor excreted by the organism. It was first discovered after Rachel Carson published her seminal book, *Silent Spring*, in 1962.[2] She noted that songbird populations in her area were rapidly declining, and she suggested a connection with the spraying of the insecticide DDT to control mosquitoes and herbivorous insects on local trees. No one knew of the mechanism of this decline at first, but then researchers noted that DDT and its derivatives increased in concentration in organisms found in higher and higher trophic levels of a given ecosystem. In a classic study of the organisms in a salt marsh in Long Island, New York, researchers found a striking increase in DDT residues of many orders of magnitude between the lower and higher trophic levels (Table 5.1).[3]

Table 5.1 shows about a thousand-fold increase in DDT residue load between the alga and the ring-billed gull. How does this happen? Since DDT and its derivatives are all fat soluble, they tend to move out of an animal's digestive tract and blood stream and be sequestered in the fatty tissues of the body. Therefore much of the ingested DDT is not excreted. These compounds are all metabolically stable, so the body cannot break them down and they remain in the fatty tissues for a long time. If a second organism ingests the first, it absorbs the DDT from the first. For example, if algal cells contain one unit of DDT, and a zooplankton eats 100 cells, the latter will then be exposed to 100 times the load in one algal cell. In reality, the zooplankton will not retain all 100 units, since some will pass through its digestive tract, but let's assume an absorption rate of 20% so the zooplankton will retain 20 units. If an insect larva eats 100 zooplankton, it will be exposed to 2,000 units (100 × 20) and retain 400 (20% of 2,000). Using the same feeding and retention rates, the next trophic level will retain (100 × 400) × 20%, or 8,000 units. You get the picture.

Table 5.1
DDT residues in organisms in a marsh in Long Island, New York.

Sample	Trophic Level	DDT Residues (ppm)
Water		0.00005
Algae	Producer	0.083
Shrimp	Herbivore	0.16
Crickets	Herbivore	0.23
Minnow	Herbivore	0.94
Black Duck	Herbivore, 1° Carnivore	1.07
Pickerel	1° & 2° Carnivore	1.33
Heron	1° & 2°+ Carnivore	3.51
Herring Gull	Carnivore, Scavenger	5.43
Osprey Egg	High Level Carnivore	13.8
Merganser	High Level Carnivore	22.8
Ring-Billed Gull	Scavenger	75.5

However, since organisms increase in size as they go up the trophic pyramid and the concentration ratios are based on animal body mass, the ppm ratios between trophic levels will be lower than those calculated in the above example. But significant magnification still does occur.

Fortunately, humans do not seem to be as sensitive to DDT as other members of natural ecosystems, but birds are very sensitive since DDT disrupts calcium deposition into eggshells. The resultant eggs have such a thin shell that the brooding adult breaks the shell, causing mortality of the developing chick. The birds most vulnerable are the top-level predators, since they have the highest DDT load. Perigrine falcon populations plummeted in the 1960's and 1970's, and when our national bird, the bald eagle, showed a similar decline, the symbolism was too great to ignore and legislated limits on DDT use were established. Even so, for a while the DDT levels in human breast milk were higher than the legal limit for the sale of commercial cow's milk, because we feed at higher trophic levels than the cow.

However, DDT is not the only compound that bioaccumulates. Dioxins and PCB's (polychlorinated biphenyl) are also fat soluble and magnify through the trophic levels. Other compounds that are not fat soluble can also bioaccumulate if they are not excreted or broken down. During the atmospheric testing of nuclear weapons, strontium 90 (^{90}Sr), a radioactive product of the nuclear fission, was released into the atmosphere. It spread to regions downwind, but at any given location it did not seem to be concentrated enough to be a threat. However, the ^{90}Sr dusted down on grass, and cattle ate the grass. Strontium is chemically similar to calcium, so the cattle concentrated it in their milk. Humans, especially children, drank the milk, and the ^{90}Sr was sequestered into the drinker's bones, adjacent to the very actively dividing bone marrow tissue. It is not a good idea to have a radioactive source concentrated near dividing cells, since radiation will produce mutations in these cells that are a major cause of cancer. So understanding the bioaccumulation dynamics of ^{90}Sr in the ecosystem helped put a stop to the atmospheric testing of nuclear bombs.

In the Adirondacks, the bioaccumulation of methyl mercury (MeHg) is a real concern for predatory waterfowl like ospreys, mergansers and loons (see Chapter 4). Mercury contamination in Adirondack lakes will be discussed more extensively in Chapter 11.

As we discuss various lake organisms in the following chapters, the terms introduced in this chapter should help in your understanding of their roles in the lake ecosystem. Habitat segregation will lead to increased species diversity, as will niche differentiation. These concepts also will be useful in understanding how human activities can impact lake ecosystems, as well as how we might mitigate these impacts.

6

THE PRODUCERS I: ALGAE

MOST SCIENTISTS AGREE THAT THE earth was formed roughly 4.5 billion years ago. During the next half billion years, the earth was bombarded with space debris left over from the consolidation of the planets, with one collision apparently severe enough to knock off an earthly chunk large enough to become our moon. It wasn't very long after that—about 3.5 billion years ago—that the first evidence of life can be found. Layered structures called stromatolites, found in Australia and dated at 3.5 billion years ago, resemble current-day mounds that are formed by bacterial colonies. (Stromatolites can also be found just south of the Adirondacks in the Petrified Gardens near Saratoga Springs, New York, but they are youngsters at a mere 500 million years old.) The complexity of the fossilized remains of a filamentous organism in Australia dated at 3.5 billion years suggests that life originated even earlier than that, perhaps as early as 3.9 billion years ago.[1]

These early fossilized organisms are made up of very simple cells, and they resemble a current group of organisms that have no nucleus or membrane-bound organelles (such as chloroplasts or mitochondria) called **prokaryotes** (*pro* ≈ before; *karyo* ≈ nut, nucleus). Because of this simplicity and their similarity to these early fossils, the prokaryotes are almost certainly the first cellular organisms on earth and have been around for more than 3.5 billion years. Present day bacteria are their current manifestation.

Early prokaryotes must have been heterotrophic, absorbing nutrients from the "primordial ooze," that organic soup from which life is hypothesized to have sprung. However, eventually some of these organisms evolved a way to trap the sun's radiant energy and store it for their own use, an early type of photosynthesis. Early such systems used organic molecules and hydrogen sulfide in these reactions, and some current organisms still use these methods. However, suitable molecules for these reactions were relatively rare on the early Earth. About 2.7 billion years ago, a photosynthetic reaction evolved using the new molecule—chlorophyll *a*—that was able to break down water, a readily available molecule, to produce oxygen and the electrons necessary to chemically make carbon dioxide (CO_2) into sugars.[2] This is essentially the same photosynthesis reaction described in Chapter 4.

Previous to this development, the earth's atmosphere contained no oxygen. In fact this oxygen-deficient atmosphere was necessary for the production of the complex organic molecules required for the initiation of life. All the existing organisms up to this time were anaerobic (capable of living without oxygen), and to most, oxygen was toxic. So the production of oxygen and its release into the atmosphere by these newly-evolved organisms produced a global catastrophe for these anaerobic organisms, and there was a massive die-off. If "pollution" is defined as "Any physical, chemical or biological alteration of air, water or land that is harmful to living organisms,"[3] the evolution of oxygen-producing photosynthesis can be considered the first global pollution episode, and it killed most of the organisms then existing on the planet.[4] Anaerobic bacteria still exist today, but they are relegated to anoxic micro-habitats, such as organic sediments and anoxic hypolimnia.

Photosynthetic bacteria have been around for so long that they have evolved a number of different ways to carry out this reaction. Sulfur bacteria use hydrogen sulfide (H_2S), and others evolved different methods and different chlorophylls. But the group that has subsequently become the most prevalent uses only chlorophyll *a*. Since this group has a blueish tint to its pigments, it is called cyanobacteria (*cyan* ≈ blue), or by the older, but I think more descriptive term, **blue-green algae**. (Throughout the remainder of this book, I will refer to this group as "blue-green algae," which is the more common usage, even though it is scientifically incorrect—they are not really algae, but photosynthetic bacteria.) They are very common in lakes today, including Sacandaga Lake and all other Adirondack lakes. We will discuss them in more detail in future sections of this chapter.

Sometime around 2 billion years ago, cells evolved into a more complex form containing membrane-bound structures such as nuclei, chloroplasts and mitochondria. These organisms are called ***eukaryotes*** (*eu* ≈ true; *karyo* ≈ nucleus), and they make up all the higher organisms we know today.

From here on, I will use scientific names when discussing many of these creatures, since common names do not exist for them. You can review the scientific taxonomic classification system in Appendix IV.

Algae

The algae are a very diverse group of photosynthetic organisms lumped together by a single characteristic—the reproductive cells of all of them are produced within a structure made up of a single cell. Aside from the blue-greens, algae are all eukaryotes, and they have a wide variety of sizes, shapes, cell wall compositions and metabolic pathways. Most are microscopic, but some marine algae can reach lengths in excess of 30 ft. In the open water of both freshwater and marine ecosystems, algae are the basis of the entire food web. In freshwater systems they make up the phytoplankton, but they can also be benthic.

Microscopic phytoplankton face a serious challenge. Although most of their cytoplasm is made up of water, many of the other cell components are comprised of molecules denser than water. As a result, algal cells tend to sink. This can be a real problem for an organism that must remain in the upper euphotic zone of a lake to survive. The rate of sinking for a small object in a fluid is governed by Stoke's Law, which relates the sinking rate to an organism's size and shape, and its density relative to the surrounding water. According to Stoke's Law, small organisms sink more slowly than larger ones, and organisms with spiky projections or elongated shapes sink more slowly than spherical ones. Organisms denser than surrounding water sink, and those less dense will float. All of these characteristics can be governed by evolutionary adaptations of the algal cells or by the metabolic adaptation of the algae in response to a stimulus.

Let's look at these characteristics one at a time. No wonder most phytoplankton are small, since smaller cells sink more slowly than larger cells.

The wide majority is microscopic, with many blue-greens the size of bacteria, about 2 microns. (A micron—μm—is 1/1,000 of a millimeter, and a dime is about one millimeter thick). Most small eukaryotes may be 5 to 10 microns in diameter. In some eutrophic lakes, there may be as many as 20 million algal cells per liter. That's 20,000 cells per mL (or a cube of water, 0.4 in on each side). A major fraction of the phytoplankton in some reaches of the ocean and most lakes had been ignored until recently since they were so small they went unnoticed. They are only 0.2 to 2 microns in diameter and are known as picoplankton. The abundance of this group in aquatic systems confirms the importance of small size in overcoming gravity in a liquid environment.

Larger phytoplankton often form filaments or have long projections from their cell walls. Such elongate and spiky shapes will increase their "form resistance" (a term in the Stoke's Law equation), thereby decreasing

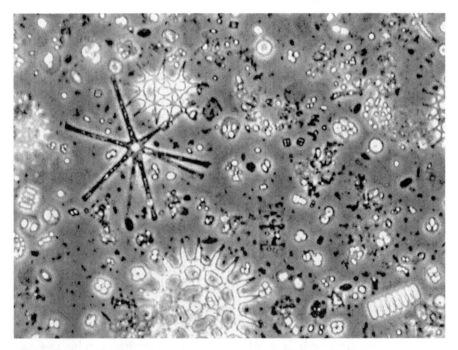

Figure 6.1
A community of algae in the eutrophic Collins Lake of Scotia, New York (magnification about 500×). Note the diversity of shapes, the small size of some, the frequency of spikes and sharp protrusions, and the filamentous structure of others. All these characteristics slow the sinking rate of the cells.

the alga's sinking rate. A sphere has a form resistance of 1.0, but for a cell with a length/width (l/w) ratio of 4, it increases to 1.3. For a l/w ratio of 15 (as in many filaments), it is 4, meaning these filaments will sink 4 times more slowly than a sphere.[5] As we will see later, these spiky shapes also tend to reduce predation from filter-feeding zooplankton.

Some algae can form gas vacuoles under certain conditions that will decrease their density relative to water. Many blue-greens will even float to the surface, forming the green scums found on many hyper-eutrophic lakes during the warmer parts of the growing season. Diatoms, which have heavy silica-rich cell walls, can form oils that are less dense than water to help compensate for the density of their cell walls. Other algae surround themselves with a gelatinous sheath, which will decrease their average density. These sheaths also protect the alga from being digested in a zooplankton's gut. Figure 6.1 shows the variety of sizes and shapes in the phytoplankton populations in a natural lake.

To illustrate the effect of size and shape on sinking rates, Table 6.1 lists some sinking rates for a variety of algae.[6] Note that the smaller of the flat, circular cells has a much slower sinking rate than the larger cell. Also, the filamentous chain of cells sinks more slowly than the single cell of the same diameter.

Blue-Green Algae (Cyanobacteria)
Blue-green algae are prokaryotic photosynthetic organisms containing chlorophyll *a*. They are quite diverse in size and shape, ranging from small spheres to large colonies visible with the naked eye. Some produce mucilaginous sheaths, others produce gas vacuoles, and one genus, *Oscillatoria*, even moves, though slowly (Figure 6.2).

One characteristic that sets the blue-greens apart from the other algae

Table 6.1
The size, shape and sinking rates of representative algae.

Cell Shape	Cell Diameter (μm)	Sinking Rate
Flat Circular Cell	2	0.2 m/day (0.7 ft/day)
	50	6.8 m/day (22.4 ft/day)
Filamentous Chain	54	1.9 m/day (6.3 ft/day)

Note. 1 micron (μm) = 1/1,000 of a millimeter.

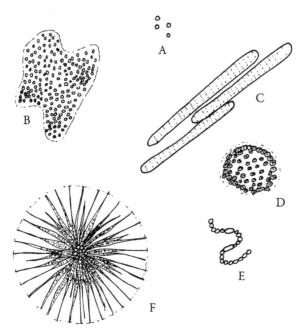

Figure 6.2
Typical blue-green algae: (A) *Chroococcus*, (B) *Microcystis*, (C) *Oscillatoria*, (D) *Coelosphaerium*, (E) *Anabaena*, (F) *Gloeotrichia*. Note that in B, D, and F, the cells are surrounded by a clear mucilage matrix. (Individual cells in A, B, D and E are 2 to 5 μm in diameter, the filament in C is 8 to 10 μm in diameter and about 100 μm long, and the colony in F is 1 to 2 mm in diameter—visible to the naked eye.)

is the ability of some of them to fix atmospheric nitrogen (N_2). Nitrogen is the most common element in the atmosphere at about 78%, but is in an extremely stable form. Only a few organisms—some heterotrophic bacteria and some blue-greens—can chemically transform this molecule into an ammonium group (NH_4^+), which can be used to make the amino acids to build proteins. This is no simple matter for the blue-greens, because they make oxygen when they photosynthesize, and oxygen poisons this nitrogen-fixing reaction. Filamentous N-fixing forms solve this quandary by producing thick-walled cells called **heterocysts**, which isolate the N-fixing reactions from the photosynthetic ones (Figure 6.3.A). If we culture potentially N-fixing filamentous blue-greens in a medium with low levels of soluble nitrogen, they will produce many heterocysts. However, some unicellular blue-greens can also fix N, and they are thought to separate these two incompatible metabolic processes temporally.[7]

Another differentiated cell among many filamentous blue-greens is the **akinete**, typically a larger, denser cell containing large stores of food and a thick wall (Figure 6.3.B). The akinete is used as a reproductive cell during periods of poor growth conditions. In one experiment, akinetes stored in a dry state for seven years were able to germinate when placed in water at a frequency of 90%.[8]

A more unpleasant aspect of the blue-greens is that some of them are capable of producing toxins. Very dense blooms like those found on warm, nutrient-rich farm ponds, can produce enough toxin to kill fish or animals drinking the water. Few if any documented human deaths have ever been reported from blue-green toxins, probably because the algal blooms must be so dense to produce sufficient toxin that the water is not palatable. The main toxins are neurotoxins and liver toxins, and some can cause skin irritation.

Blue-greens tend to reproduce more rapidly at warmer temperatures and are rarely obvious in our lakes early in the growing season. However, after the fall turnover when the higher phosphate levels of the hypolimnion are mixed into the euphotic zone, blue-greens may become more prevalent. If the blue-green species present at that time are capable of forming gas vacuoles, they will rise to the surface and may collect on the downwind shore of the lake, forming a localized blue-green slick. Even though you cannot be sure that the slick is formed by a toxin-producer, it is probably not a good idea to swim in such water. If you must get into the water, for instance to take in a dock, take the precaution to wash off your exposed skin areas with soapy water.

Our Adirondack lakes have not seen the main toxic forms of blue-green algae—certain species of *Anabaena*, *Aphanizomenon*, *Microcystis*, affectionately called "Annie, Fanny and Mike" by those who work with them. But we have another blue-green, which, in dense blooms, can cause problems. *Gleotrichia echinulata* is a common alga found in many clear lakes (Figure

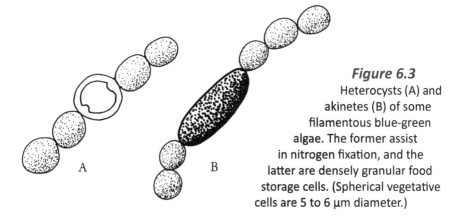

Figure 6.3
Heterocysts (A) and akinetes (B) of some filamentous blue-green algae. The former assist in nitrogen fixation, and the latter are densely granular food storage cells. (Spherical vegetative cells are 5 to 6 μm diameter.)

6.2.F). It germinates its akinete on the sediment surface, grows down there by extracting nutrients from the sediment, then produces gas in its vacuoles so that it becomes suspended in the water column. Its colony is covered with a spherical gelatinous sheath that reaches a diameter of 1 to 2 mm, so it is visible with the naked eye. It appears like a translucent dot in the water, and it is quite obvious when present. Moderate blooms have been described at 400 colonies per liter,[9] but I have never seen more than about 20/L in Sacandaga. Therefore, we are nowhere near a dense bloom, the level at which problems occur. Its presence in our lakes does not indicate eutrophic conditions, however, since it gets most of its nutrients from the sediment.

Clinical studies using dermatological patches have shown that many blooms of known toxin-producing blue-greens do not produce toxins, and only a small proportion of the human population (3 to 20% depending on the study) is sensitive to a given blue-green skin irritant.[10] So even though this group of algae can be problematic in some cases, they do not seem to be a problem (yet) in the Adirondacks.

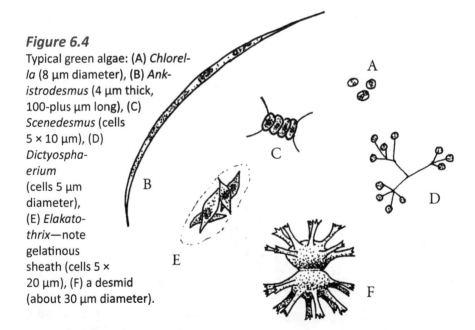

Figure 6.4
Typical green algae: (A) *Chlorella* (8 µm diameter), (B) *Ankistrodesmus* (4 µm thick, 100-plus µm long), (C) *Scenedesmus* (cells 5 × 10 µm), (D) *Dictyosphaerium* (cells 5 µm diameter), (E) *Elakatothrix*—note gelatinous sheath (cells 5 × 20 µm), (F) a desmid (about 30 µm diameter).

Green Algae

Green algae are eukaryotic and contain two chlorophylls (*a* and *b*). These two chlorophylls and many other biochemical and morphological characteristics can be found in all higher plants, suggesting that the green algae are ancestors to all higher plants we know today.

Green algae are also quite diverse. Most are planktonic, but many are filamentous and benthic. Their structure is more complex than the blue-greens, and some of their cell walls can be quite ornamented. Stokes Law also governs their sinking rates, so small size, elongated shape and long cell wall projections are very common (Figure 6.4). A few have flagella and can swim to compensate for their normal sinking rate.

Diatoms

Diatoms are typically unicellular planktonic and benthic eukaryotic algae containing chlorophylls *a* and *c*, with accessory pigments that are yellow to brown (mostly different from the green algae).[11] They store their food as oils. However, the most distinctive characteristic of the diatoms is their cell walls, which are made up of rigid silicate and consist of two halves that fit together with overlapping edges like a pillbox. They are also delicately ornamented (Figure 6.5). It is important for a cell with such rigid and impermeable cell walls to have holes in the walls so that the living cell inside can receive nutrients from its environment, but why those "holes" are so symmetrically placed and different in every species is a mystery. We can just enjoy their beauty.

Diatoms are very common in both freshwater and marine habitats. Their siliceous cell walls make them heavier than other planktonic algae, and they at least partially compensate for this by producing low-density oil as a food storage product. In spite of this, their sinking rate is usually fairly high, and they depend on water turbulence to remain in the euphotic zone. Marine diatoms have been shown to take up luxuriant levels of nitrates when available to maintain a high photosynthetic rate. If freshwater species can do the same, they may contribute to the precipitous drop in nitrates in the epilimnion of Sacandaga Lake in the spring (see Figure 4.4) due to their relatively rapid sinking rates. They also strip silica out of the epilimnion, and the lack of these two nutrients most likely limits diatom growth later in the growing season, so they are much more prevalent in the spring and early summer.

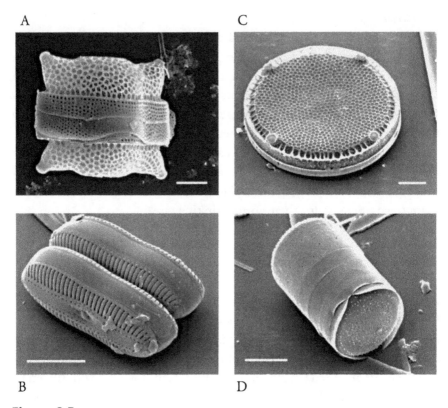

Figure 6.5
Scanning electron microscope images of diatoms, showing the intricate patterning in the silica cell walls.[21] The top and bottom halves of the cell wall fit together like a pillbox. The "girdle" in the middle of the cell is where the two halves overlap. (A) *Biddulphia*, (B) *Diploneis*, (C) *Eupodiscus*, and (D) *Melosira*. (White bar is 10 μm long.)

Diatoms are also an important component in the benthos. The brownish "scum" that you see covering rocks in streams and lakes is predominantly diatoms. In fact, on a sunny day in a shallow stream, you can often see small gas bubbles on these scummy rocks. Those bubbles are the oxygen produced by the diatoms' photosynthesis. Diatoms are equally important on the surfaces of the rocks and sediment in lakes, and are a major source of food for the complex community of animals in the "scum." Some of these benthic diatoms are even mobile, pulling themselves along on fibrils extending from a slot in their walls.

Other Algal Groups

Another alga common to Sacandaga and other Adirondack lakes is *Dinobryon*, a colonial alga that looks like a tiny tree. Each cell has a vase-shaped, transparent lorica (a hard, shell-like structure), all of which are attached in a minute arboreal configuration (Figure 6.6.C). The cells have two flagella, so they can be mobile. *Dinobryon* is known to survive well under conditions of low dissolved phosphorus.[12] A possible reason for this is that they recently have been shown to be both autotrophic (photosynthetic) *and* heterotrophic (able to ingest bacteria). Under certain circumstances, these ingested bacteria can contribute up to 77% of the total phosphorus requirement for this alga.[13]

From the above discussion, you can see that the definition of *Dinobryon* as an "alga" is somewhat questionable due to its capacity to ingest bacteria. In fact it even blurs the distinction between plants and animals. The next group pushes that envelope even more. The dinoflagellates are a diverse group of unicellular algae with chlorophylls *a* and *c*, similar to the diatoms, but brownish accessory pigments with food stored as starch. They

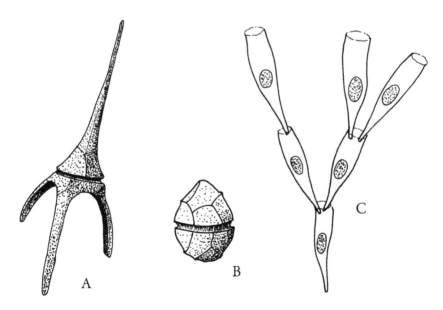

Figure 6.6
The dinoflagellates *Ceratium* (A) and *Peridinium* (B), and *Dinobryon* (C). All have flagella (not shown) and are mobile. (*Peridinium* is 50 µm long, and others are to same scale.)

can also have plate-like cell walls that remind one of armor, and a groove around their midsection containing a flagellum. Another flagellum extends out from the cells, making them very mobile. They also have weird bead-like chromosomes and a unique process of cell division.

There are two dinoflagellates common to our lakes—*Ceratium* and *Peridinium*. Our species of *Ceratium* (*hirundinella*) has four long spines extending out from the main cell body, three on the "bottom" and one on the "top" (Figure 6.6.A). It reminds me a little of a spaceship. The spines are useful in slowing the cell's sinking rate as well as deterring predatory filter feeders. The "armor" plates are quite obvious, as is the central groove. *Peridinium* is more oval, with no extensions but obvious plates and groove (Figure 6.6.B).

Ceratium is very common world wide, and it is relatively large with a distinctive shape, so a lot of research has focused on it. For instance, micro-analytical chemical analysis of *Ceratium* cells show that it can concentrate phosphorus in its cytoplasm to levels 10,000 times greater than that of the surrounding water.[14] But the characteristic that attracts many researchers is its ability to move within the water column on a daily basis. Many aquatic animals show this diurnal movement, typically rising to the surface during the nighttime and swimming deeper during the daylight hours. But *Ceratium* is one of the few unicellular organisms that are strong enough swimmers to show measurable migration. It has been shown to migrate up and down so that it aggregates at an irradiation level of about 10% of the surface light intensity.[15, 16] These migrations can be on the order of about 10 ft (3 m) each way. However, for these migrations to occur, the water column must be very stable under calm wind conditions. During windy conditions in which the epilimnion is well stirred, the populations are evenly dispersed.[17]

There are a few hypotheses regarding the adaptive advantage of these diurnal vertical migrations. Migrations occur in many microscopic aquatic organisms, so there must be some advantage, given the amount of energy expended by such small organisms during their 20 ft (6 m) daily round trip migrating through what, to them, is a very viscous medium. One hypothesis suggests that the reason lies in the difference in chemical composition of the epilimnion and the hypolimnion. By spending at least some time in the relatively nutrient-rich hypolimnion, an organism can absorb valuable limiting nutrients that may be rare in the epilimnion. Another idea suggests an optimum illumination level for photosynthetic organisms, such as the 10% surface illumination level mentioned above. Photosynthetic inhibition can occur at radiation levels that are too high. A third hypothesis pertains to the vulnerability of prey to visual predators. During the day, prey species swim down to levels of illumination where it is difficult for the predators to see them. They can then rise to the warmer, more oxygenated waters of the epilimnion during the darker hours.[18] Probably all of these hypotheses are valid to some degree.

Perhaps one of the more bizarre characteristics of the dinoflagellates—fortunately not the ones we find in the Adirondacks—is that many of them are toxic. Marine dinoflagellates cause the red tides that produce the neurotoxins shellfish concentrate, making the shellfish poisonous to humans. But one of the most problematic is *Pfiesteria piscicida*, sometimes called the "cell from hell."[19] At last count, researchers had found 24 different stages in its life cycle, some of which are photosynthetic while others heterotrophic. It has been associated with massive fish kills in estuaries from North Carolina to Maryland, and one of its stages even attacks fish with a little bayonet-like projection. During an intense bloom, *Pfiesteria* can produce a neurotoxin that can become airborne and affect humans merely inhaling the air near a bloom. Symptoms of exposure in humans include headaches, rashes, confusion, and even loss of short-term memory for months if exposure is severe enough. Thank goodness our Adirondack dinoflagelates are more on the mellow side.

"Macro-Algae"

There is one group of plants that at one time was included with the algae due to similar reproductive structures. Although simple in structure—in some species its "stem" consists of a single elongate cell—they are now grouped with the higher plants, though they are still often called by their older name "macro-algae." No matter where you place them taxonomically, they are common in most Adirondack lakes and can produce a carpet on the sediment at the lower edge of the littoral zone.

They are benthic and large for "algae"—up to 12 in (30 cm)—and can live at depths vascular plants cannot. For instance in Lake George, the deepest-growing vascular plant is the common waterweed, *Elodea canadensis*, at 39 ft (12 m), yet there is a "meadow" of the macro-alga *Nitella* living from about 40 to 50 ft (12 to 15 m).[20] The light intensity at 39 ft in Lake George is just above the 1%, considered the lower limit to the euphotic zone. *Nitella* grows nicely in waters with low calcium levels, but a related macro-alga, *Chara*, requires higher calcium levels. *Chara* precipitates calcium as calcium carbonate (like limestone) on the periphery of its stems, which gives it a brittle texture, hence its common name "stonewort." Its aroma when pulled from the water leads some to name it "muskgrass." It can grow in some Adirondack lakes with sufficient calcium levels (Figure 6.7).

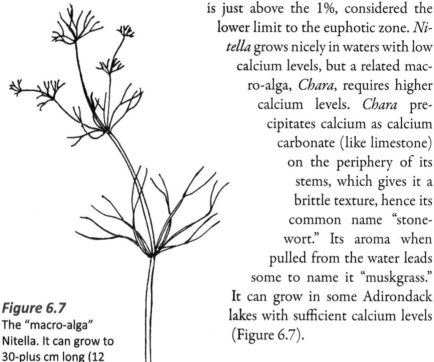

Figure 6.7
The "macro-alga" Nitella. It can grow to 30-plus cm long (12 in), and the stem-like structure between the clusters is a single long cell.

Algal Productivity in a Lake

When we speak of algal "productivity," we really mean the amount of photosynthetic product (sugars and other organic compounds) produced by the autotrophs in a community. From Chapter 4, you will recall the equation for photosynthesis:

$$\text{Carbon Dioxide} + \text{Water} + \text{Light Energy} \rightarrow \text{Oxygen} + \text{Sugar (Glucose)}$$

Since oxygen is a product of this reaction, we can measure the photosynthetic rate by measuring the rate of production of oxygen in the water column. However, all living organisms, both plants and animals, respire, using the sugar and other organic compounds produced by the autotrophs plus oxygen to produce the metabolic energy needed to live. So respiration uses oxygen and photosynthesis produces it. It would be nice if we could separate these two processes to understand what is really going on

The Secret Life of a Lake

in a lake.

We can do this by altering the light available to the biological community in the lake. If light is excluded, only respiration (RS) will be occurring. If light is available, both RS and photosynthesis (PS) will be occurring. If we take a sample of lake water containing the biological community and seal it in a bottle for a few days, the organisms in the bottle will alter the dissolved oxygen (DO) levels in the bottle by both RS and PS. If the bottle is exposed to light, both PS and RS will occur, but if the bottle is covered with opaque tape, only RS will occur. This experiment, cleverly called a "light bottle/dark bottle experiment," allows us to determine how much PS is going on at various depths in the lake, as well as estimate the biomass (total mass of all the organisms) present along the sampled depth profile by determining the organisms' respiration rate.

To run this experiment, we collect water samples from various depths in the lake and use water from each of these depths to fill three bottles—a light bottle, a dark bottle, and a third bottle used to determine the oxygen levels at that depth at the start of the experiment. We suspend the light and dark bottles in the lake on a calibrated rope so they rest at the same level that we collected the water sample, from the surface down well into the hypolimnion (Figure 6.8). We allow the series of bottles to remain in the lake for a few days, then collect them and measure the dissolved oxygen (DO) lev-

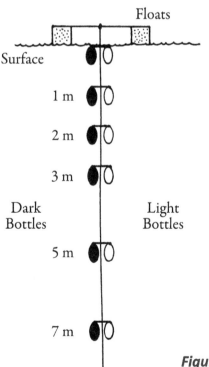

Figure 6.8
A diagram of the setup of the light/dark bottle experiment to measure a productivity profile.

els in each, and compare them to the DO at that depth at the start of the experiment. We can therefore calculate the oxygen increase or decrease in the light and dark bottles over the time period of the experiment.

In the dark bottles, only respiration (RS) will occur, so the DO will decrease relative to the initial DO levels. The amount of decrease in DO relates to the total biomass of the organisms at that sample depth. We usually find that there are more organisms in the top of the epilimnion, and their biomass decreases slightly with depth. There may be a peak at the top of the thermocline, indicating the group of organisms that can take advantage of the water density gradient provided by the thermocline to remain suspended in the water column. In the hypolimnion, biomass decreases significantly, but there is always some present and DO levels always decline.

In the light bottles, both photosynthesis (PS) and respiration (RS) occur, and appropriate calculations can separate the PS from the RS. Photosynthesis is related to both the light intensity and the biomass of algae present at a given depth. We would therefore expect that PS would be greatest at the surface, but this is rarely the case because high light intensity can actually *inhibit* PS at the surface. Through the remainder of the epilimnion, PS typically decreases with depth, as light intensity decreases. If the thermocline is relatively shallow and well into the euphotic zone, as often occurs in the beginning of the summer, we sometimes see a PS maximum indicating a layer of high algal biomass suspended in the density gradient at the top of the thermocline. (Remember the DO peak in the June curve in Figure 4.2?)

As the light intensity decreases farther into the hypolimnion, PS will at some point equal RS, so there will be no net change in DO in the light bottle. This point where PS equals RS is called the ***compensation point***, and it is usually defined as the lower limit to the euphotic zone. Below the compensation point, dissolved oxygen levels will decline over the growing season, and this is the reason for the decrease in hypolimnetic DO after the spring turnover. The rate of decline depends on water transparency, which describes how deeply light can penetrate into the lake, and the amount of organic material available to feed the heterotrophs in the hypolimnion. In a eutrophic lake with high nutrient levels, the high biomass in the epilimnion prevents light from penetrating very deeply, and there is considerable organic material for the heterotrophs to feed on and metabolize, so the euphotic zone is very shallow. The low levels of PS in the hypolimnion and high levels of RS conspire to deplete hypolimnion DO levels very quickly, and they remain low until fall turnover.

In an oligotrophic lake, where algal biomass is limited by low nutrient amounts, light can penetrate to very deep levels (thereby increasing PS), and the low algal biomass provides only a little organic material for the heterotrophs to eat and metabolize (thereby decreasing RS). The result allows the hypolimnion to maintain high levels of dissolved oxygen throughout the entire summer. Sacandaga Lake, which is classified as meso-oligotrophic, falls somewhere in between, with its hypolimnetic DO being depleted for only a short time at the end of the summer.

So the algae are a major component of the pelagic (deep water) biological community in a lake. They are the basis of the food chain, providing nourishment for the microscopic animals that feed on them. They are a main driver of the oxygen levels in most lakes, and intense blooms can even produce fish kills by decreasing the oxygen levels as they decompose. In certain situations, fortunately not in the Adirondacks, they can even produce toxins that can affect other creatures in the lake, and even humans in rare extreme cases.

Algae are also incredibly diverse, with their various biochemical pathways suggesting complex evolutionary origins. Some are mobile, swimming to regions of the lake where they find optimal conditions for growth. Some migrate up and down the water column on a diurnal basis. Others just hang out, drifting with the small currents in the mixing epilimnion. Sooner or later almost all settle down into the hypolimnion where they are decomposed by resident bacteria, their nutrients eventually recycling during turnovers up into the euphotic zone where they are incorporated into new algal cells.

7

THE PRODUCERS II: VASCULAR PLANTS OR "WEEDS"

ALGAE ARE THE BASIS OF the open water food chain, but vascular aquatic plants rule the shallower reaches of a lake. The term "vascular plant" simply refers to the fact that they contain specialized transport tissues made up of the vascular tissues xylem and phloem. They are evolutionarily more advanced than even the most complex alga. They all have chlorophylls *a* and *b*, carotenes as accessory pigments, cellulose cell walls, and store their food as starch. These characteristics lead us to the inescapable conclusion that they all evolved from green algae, the group that also contains these compounds. The term "weed" is a subjective description of a plant that is in some way undesirable, and it is commonly used to describe aquatic plants. I hope that after reading this chapter you will gain some real appreciation for these botanical outcasts.

There are some intermediate groups between the algae and vascular plants, the mosses and other bryophytes, that are more advanced than the algae but have no xylem or phloem. These groups are common in some stream habitats, but are relatively rare in Adirondack lakes and will only be briefly mentioned here. Vascular plants most likely evolved from terrestrial mosses and later returned to the aquatic habitat, as suggested by the anatomy of their vascular tissue. To appreciate the adaptations they had to develop to make this change, we must first discuss the plant characteristics required to survive the desiccating terrestrial environment.

Mosses have no vascular tissue or roots and they have very simple leaves, some of which are only one cell thick. Terrestrial mosses do not avoid drought, they endure it, but while doing so they are metabolically inactive. Some desert mosses are metabolically active only the few weeks a year when they are moist. Plant groups more advanced than mosses developed thicker leaves covered with a modified epidermis protected by a waxy, waterproof layer called a cuticle. They also developed roots to penetrate the substrate (which is almost always wetter than the ground surface), vascular tissue to transport water and nutrients among the roots, stems and leaves, and strong support tissue to allow the leaves to reach higher light intensities. With these adaptations, a terrestrial plant could now survive moderate drought conditions while remaining metabolically active most of the year.

Aquatic Adaptations

When these terrestrial plants expanded their ranges into aquatic habitats, many of these adaptations that were so beneficial in the terrestrial habitat became problematic in the aquatic environment. Aquatic plants that are entirely submerged have no need for strong support tissue, since the water buoys up tissues containing even a minimal amount of gas. Likewise, the thick, waterproof leaf impedes gas transport into and out of the leaf, so the cuticle was lost and leaves became thin once more. Since many nutrients can be absorbed directly from the lake water, vascular tissue was reduced substantially. Emergent aquatic plants with some plant parts exposed to the air retained the thick leaves and support tissue in the parts of the plant rising above the water level.[1]

Lacunae

But the adaptation typical of many aquatic plants that is not found among their terrestrial relatives is the large openings found in their stems called *lacunae* (Figure 7.1). These lacunae are continuous from the leaves, through the stems to the roots, and they are filled with gasses. They act like transport tissue for gasses in the plant, since gasses diffuse about 10,000 times faster in a gaseous state than they do when dissolved in a liquid. Many aquatic plants have their roots embedded in organic sediments that, due to the bacterial decomposition in the sediment, are typically devoid of oxygen. Roots are alive and metabolically active and therefore need oxygen to survive, and the lacunae are able to supply it by simple diffusion.

One of the more bizarre findings about aquatic plants relates to these lacunae. In the yellow water lily (*Nuphar*), which is common in our Adirondack lakes, researchers found that its older leaves emitted methane gas during the day, but its younger leaves did not. Since methane (CH_4) is a product of anaerobic decomposition of organic matter, it is very common in marshy areas. (When you walk through a marshy area, the bubbles that arise from the sediment are mostly methane.) It is also a greenhouse gas, so the researchers tried to understand the role of the lily relevant to the atmospheric methane budget. What they found was astonishing—that the lily actually produced "internal winds" in its lacunae, originating at the younger leaves and "blowing" toward the older leaves.[2,3] Using radioactive and other tracers, they found that "wind" velocities in a ***petiole*** (the stem-like structure at the base of a leaf) can reach 50 cm/min (18 in/min) and a volume of 22 L/day can flow through a single petiole. When the young leaves get warmed by the sun to above ambient temperatures, a combination of the thermal expansion of gasses and increased evaporation inside the leaf produces a slight positive pressure. This pressure pushes the air in the lacunae down the young leaf's petiole, into the rhizome (a submerged modified stem) and back up the petioles of the older leaves, which are far more porous that the younger leaves. The rhizome is embedded in the anoxic organic sediment and therefore supplies the methane. These "winds" from the young leaves also supply the rhizome with ample oxygen to support its metabolism. During the night, thermal equilibrium is reached and the "winds" stop. And to think all this is going on in those lovely yellow water lilies, so common in our lakes.

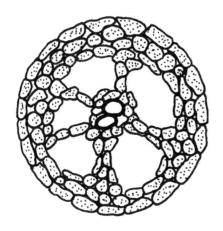

Figure 7.1
The lacunae in the stem of an aquatic plant. Note the five large air spaces that are filled with gasses. These give the plant buoyancy, as well as provide channels for gas transport among the leaves, stem and roots.

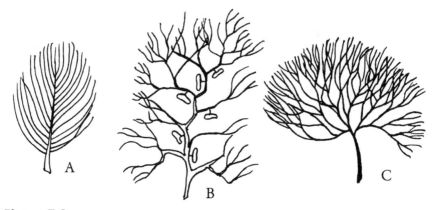

Figure 7.2
Feathery leaf morphology that enhances absorption of nutrients from the water. Leaves are from (A) a milfoil, (B) a bladderwort, *Utricularia* (note traps), and (C) fanwort, *Cabomba*. All leaves are about 5 to 8 cm long.

Surviving Anoxia

Another adaptation of aquatic plants rooted in anoxic sediments is their ability to metabolically tolerate extremely low oxygen levels. Respiration in the absence of oxygen is called fermentation, and its end product is ethyl alcohol (ETOH). This is the same process we use to make beer and wine. Eventually the ETOH concentrations rise to lethal levels, unless the plant has evolved a means to either bypass the ETOH production or break it down before it reaches a lethal state. White water lilies (*Nymphaea*), cattails (*Typha*), and pickerel weed (*Pontederia*) survive anoxic root conditions by keeping ETOH levels low, and the marine marsh grass (*Spartina*) has been shown to increase its alcohol dehydrogenase (an enzyme which breaks down ETOH) levels fivefold in roots growing in anoxic organic sediments.[4]

Nutrient Acquisition

Aquatic plants can get their nutrients from both the sediment and the lake water. Nutrients are far more concentrated in the sediment and that is the source for most plants, assuming their roots can survive in the sometimes-anoxic sediment.[5] Sandy sediments hold fewer nutrients than organic sediments and their oxygen levels are usually higher. Since nutrient absorption is a surface phenomenon, some aquatic plants have evolved submerged leaves with a highly dissected, feathery morphology, in essence modifying a flat laminar leaf into a series of cylindrical tubes. This enhances their ability to absorb nutrients directly from the water (Figure 7.2). When you compare the surface/volume ratio (S/V) of a flat laminar leaf with a feathery leaf with tubes the same diameter as the thickness of the laminar leaf, this ratio will increase 2 times, even though the volumes of the two leaves are identical.[6] Some plants produce thin or feathery leaves under water and thicker, "terrestrial" leaves on or above the water's surface, a characteristic known as heterophylly (*hetero* ≈ different; *phyll* ≈ leaf).

A novel approach to nutrient acquisition by an aquatic plant is shown by the common bladderwort (*Utricularia vulgaris*), a frequently found plant in our Adirondack lakes. It has no roots, so it does not have access to the nutrient-rich sediment. This plant has feathery leaves (Figure 7.2.B) to increase its S/V ratio, but it also has modifications on its leaves in the form of small pouches—the "bladders" of the bladderwort (Figure 7.3).

These bladders have a very complex anatomy and physiology, and they are capable of catching small aquatic animals such as insect larvae and large zooplankton. Therefore the bladderwort can be considered a carnivorous plant. The bladders are surrounded by a thin membranous wall that is one or two cells thick. There is a "door" that can make a tight seal with the bladder opening, and some species of *Utricularia* have long filaments

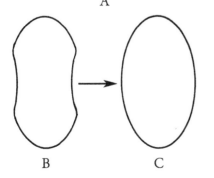

Figure 7.3
The trap of *Utricularia* (2 to 3 mm). Cross sections B and C are from position A–A on the trap. In B the trap is set, and in C it has been tripped. D shows the door that contains the small filaments (the two small ones at the bottom of the door) that spring the trap. (This illustration has been modified from the original.[7])

extending from the opening that are thought to guide unwary prey towards the door. The door itself has many smaller bristles, which when contacted by a "prey," will deform the door so that the seal is broken and the door can open. The trap is set when cells in the wall actively secrete ions out of the bladder so that water osmotically follows the ions from the bladder into the surrounding water. This leaves a slight negative pressure inside the bladder, and the walls are drawn in as shown in Figure 7.3.B. When a prey touches the bristles on the door, the door seal releases, the door opens and the springy bladder walls expand outward, increasing the volume inside the bladder and thereby sucking the prey inside (Figure 7.3.C).[7] The door then closes, and glands on the inner side of the bladder wall secrete enzymes that digest the prey. Labeled phosphorus and nitrogen atoms incorporated into prey specimens are taken up by the *Utricularia* plant tissues within a few days.[8] It takes only a few milliseconds for the trap to spring, but if the traps are sprung by something other than a prey (such as water movement), they can reset within 40 minutes. If you lift a *Utricularia* plant out of the water, you can actually hear the bladders pop open with a slight crackling sound.[9]

Some researchers have questioned the utility of the bladders as traps since only about 10% of the traps contain prey in the natural habitat, and they suggest instead that there is a beneficial mutualistic symbiotic relationship between the plants and the biological communities living inside the bladders that is more important to the plant than carnivory. Whatever the primary role of the bladders, the plant expends a lot of energy producing and maintaining them—in some species bladders make up 50% of the total biomass—and *Utricularia* is a very successful genus of plants. There are over 200 species worldwide, and they make up over 40% of the total number of plant species considered "carnivorous."[10]

A Photosynthetic Adaptation

Another adaptation of aquatic plants compared with terrestrial plants is their ability to use the bicarbonate ion (HCO_3^-) as a source of carbon for photosynthesis. Terrestrial plants typically use only CO_2 in their photosynthetic reactions. To use HCO_3^- requires a few extra enzymes and reactions. However, CO_2 is not readily available in many aquatic situations because when CO_2 is dissolved in water, the following reactions occur:

$$(CO_2 + H_2O) \leftrightarrow (H_2CO_3) \leftrightarrow (H^+ + HCO_3^-) \leftrightarrow (2\,H^+ + CO_3^-)$$

Acidic Water ←→ Neutral Water ←→ Basic Water

All these reactions are reversible, and they are driven to the left in acidic water and to the right in more neutral or basic water. At acidities in the neutral range, most of the dissolved carbon is in the form of bicarbonate (HCO_3^-), in some cases many hundred times more concentrated than carbon dioxide.[11] So those plants that have evolved the capability of using bicarbonate have a selective advantage in neutral to basic waters, but those living in acidic lakes can retain the original carbon dioxide pathways. At least one plant that uses only CO_2 for photosynthesis, the water lobelia (*Lobelia dortmanna*), has evolved an unusual way to extend its range into more neutral and basic lakes where CO_2 is rare. Its photosynthetic leaves form a small basal rosette close to the sediment, and it uses the large lacunae in its leaves to transport CO_2 from the sediment up to its leaves where it is fixed by photosynthesis (Figure 7.4.F).[12] This species of *Lobelia* is common in our lakes.

Limiting Factors

So what determines which aquatic plants will grow where in a lake? The main limiting factor is water depth. Emergent plants cannot grow at very great depths, but even some plants that spend most of their life cycle submerged require access to the surface to reproduce. An indirect effect of depth is light intensity, which is dependent on water clarity. In the very clear Lake George, the common waterweed (*Elodea*), a rooted vascular plant, can live as deep as 12 m (about 40 ft).[13] In Sacandaga Lake, and most other Adirondack lakes that are not as clear as Lake George, the lower limit of vascular plants is much shallower.

Sediment characteristics also limit the ranges of plants. Sandy sediments are low in nutrients but contain more oxygen, whereas the reverse is true for heavily organic sediments. Wave action on downwind shores tends to produce sandier substrates, and calm bays have more organic sediments. A plant's ability to tolerate low oxygen levels in the sediment will also determine where it can be found in a lake. And a plant's mechanisms for using bicarbonate versus carbon dioxide for photosynthesis will determine its pH range. Therefore the subtle differences in various plants' metabolic

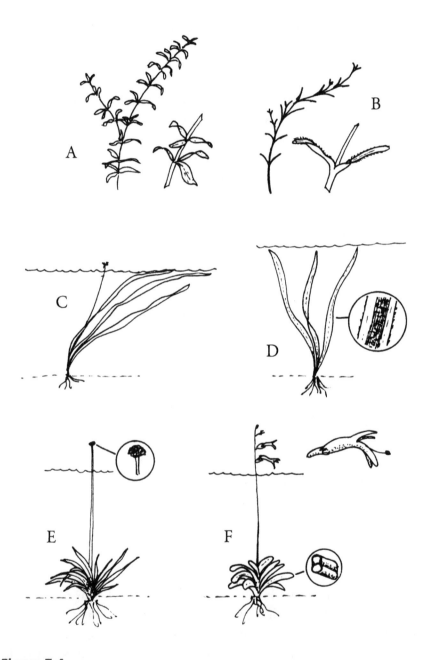

Figure 7.4
Aquatic plants with submerged leaves, with identifying details: (A) common waterweed, *Elodea*, (B) slender naiad, *Najas*, (C) bur reed, *Sparganium*, (D) water celery, *Valisneria*, (E) pipewort, *Eriocaulon*, (F) water lobelia, *Lobelia*.

tolerances to these many variables will ultimately determine where a given species will prosper.

Vascular Plants

In the next sections of this chapter we will discuss a few of the vascular plants common to Adirondack lakes, starting at the deeper reaches of the littoral zone. As a reminder of the different depth zones in a lake, you may wish to refer back to Figure 5.1.

Submerged Plants

Many aquatic plants can survive in deeper water, limited only by the intensity of light. To maintain a sustainable population, they must be able to reproduce underwater, through asexual fragmentation or by extending a flowering stalk to the surface for pollination. The common waterweed (*Elodea canadensis*) is often used as an aquarium plant and is familiar to most of us (Figure 7.4.A). It is short in stature, rarely growing to 1 m in length, and can form dense stands. It can tolerate very low light conditions and, as mentioned previously, is the deepest growing vascular plant in Lake George. In the slender naiad (*Najas flexilis*), pollination takes place under water (Figure 7.4.B). It is a small, bushy plant that can achieve high densities and live at deeper depths than plants requiring pollination above the water surface. A small, leafless and inconspicuous native dwarf water milfoil (*Myriophyllum tenellum*) reproduces using asexual budding, so does not need to reach the surface for reproduction.

Other plants have submerged leaves, but produce flowers that rise above the water surface. Wild celery (*Vallisneria americana*, which really doesn't look much like celery) has long flat leaves, about a half inch wide, and it grows in water about 3 ft deep. Its leaves have prominent vein structure running up the middle of the leaf with characteristic cross-hatching (Figure 7.4.D). This plant has tiny 1 mm (1/25th in) male flowers that are released from the plant, and they float to the surface by means of a small air bubble enclosed in their buds. Once on the surface, they "sail" around, blown by the breezes. The female flower extends to the surface attached to a long, spiral stem. Once the female flower is pollinated, the stem contracts its spiral, drawing the fertilized flower beneath the surface and away from surface-feeding herbivores such as ducks.[14]

In some of the pondweeds (*Potamogeton* sp.), only the flowers rise above the surface, but most of our pondweeds have at least one floating leaf. Two plants growing in Sacandaga Lake that are often confused have small rosettes of leaves, 4 to 8 in long, that grow down on the substrate. They both grow in 2 to 4 ft of water and extend a flowering stalk up above the surface. In the pipewort (*Eriocaulon aquaticum*, Figure 7.4.E) the stalk is terminated by a small bulbous structure containing inconspicuous flowers, and its leaves are sharply pointed. In the water lobelia (*Lobelia dortmanna*, Figure 7.4.F) there is a spike of small, one half inch, whitish flowers. Its leaves are blunt and tubular, each containing two large lacunae (see detail in 7.4.F). Species of the bur reed (*Sparganium* sp., Figure 7.4.C) have long, thin grass-like leaves. In some species, the leaves remain below the water surface, while in others the tips of the leaves extend up and float along the surface. A small native milfoil called Farwell's milfoil (*Myriophyllum farwellii*) can be found in some of the shallow, calm parts of our lakes, as can the bladderwort (*Utricularia vulgaris*) mentioned previously for its role as a carnivore. Both are often confused with their nasty invasive relative, the Eurasian water milfoil.

Plants with Floating Leaves

Many aquatic plants have both submerged and floating leaves while others have only the latter. The large genus *Potamogeton*, or the "pondweeds," has many species with mostly submerged leaves, but only a few—sometimes only one—floating leaf. These leaves are usually shaped like an elongated oval, less than 4 in long, and their upper surfaces are so waxy and waterproof that surface water drops bead up and roll off. The large-leaved pondweed (P. *amplifolius*, Figure 7.5.A) can grow in fairly deep water, 6 to 10 ft, but typically has at least a few oval floating leaves of 2 to 4 in long and 1 to 2 in wide. Its submerged leaves can grow up to 8 in long and 3 in wide, and they are often curved backward into a sort of banana shape. This plant has the largest leaves of any of our aquatic plants. Other species of *Potamogeton* may have linear, grass-like submerged leaves (P. *natans*, P. *epihydrous*, Figure 7.5.B), or broader leaves (P. *perfoliatus*), but they all have oval floating leaves.

But the floating leaves we are most familiar with are those of the water lilies. We have two water lilies in Sacandaga Lake, the white (*Nymphaea odorata*, Figure 7.5.C) and yellow (*Nuphar variegata*, Figure 7.5.D). When they are flowering, the difference is obvious. The white lily flower has nu-

The Secret Life of a Lake

merous beautiful white petals with many yellow stamens in the center. The configuration of this flower and the broad petal-like stamens in the center show that *Nymphaea* is an evolutionarily primitive genus—closely related to some of the earliest known flowering plants. In contrast, the flower of the yellow water lily is quite different, with just 5 to 6 yellow sepals that look like petals that are usually tightly appressed to the internal flower parts. The

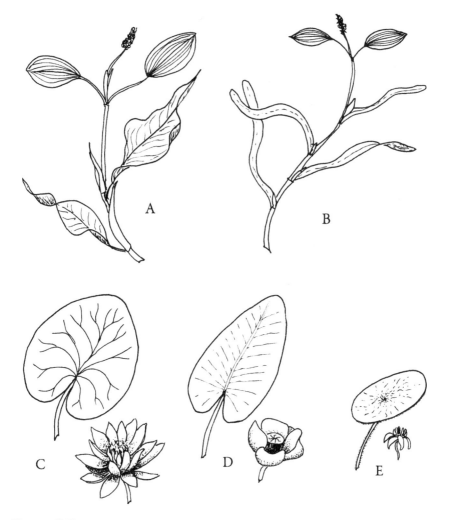

Figure 7.5
Plants with floating leaves: (A) large-leaf pondweed, *Potamogeton amplifolius*, (B) ribbon-leaf pondweed, *P. epihydrous*, (C) white water lily, *Nymphaea*, (D) yellow water lily, *Nuphar*, (E) water shield, *Brasenia* (dotted edges represent slime layer).

floating leaves of these two lilies are a little more confusing to differentiate. The leaves of the white lily are more roundish, with a cleft where the petiole (leaf stalk) originates, whereas in the yellow lily the leaves are more oval with a prominent central midrib and side veins spreading in a parallel pattern from the midrib to the leaf edge. Again, the petiole originates in a cleft in the leaf. To make matters even more confusing, a relative newcomer to our lakes looks like a smaller version of *Nymphaea*, and is called *Nymphoides*. Whereas the former has leaves typically 4 to 6 in wide, the latter's leaves widths are usually less than 4 in.

Another common plant with floating leaves is the water shield (*Brasenia schreberi*, Figure 7.5.E). Its leaves are oval, about 4 in at their long dimension. However unlike the other two species discussed above, *Brasenia's* leaves have their petiole originating at the center of the leaf. In addition, the lower portion of the leaves and the petioles are usually covered with a gelatinous layer, giving them a slimy texture. You might wonder about the utility of this slime, given the enormous amount of energy the plant expends to produce it. A reasonable hypothesis might suggest predator avoidance, but studies have showed that an aquatic caterpillar that can frequently be found eating water lilies preferred *Brasenia* and gained more weight from eating it when compared with other aquatic plants. In fact, when researchers spread various aquatic plant substances on willow leaves, the caterpillar preferred the leaves coated with *Brasenia* slime to the other experimental applications.[15] One more brilliant hypothesis shot down by a simple experiment. At this point, I'm not sure anyone knows the function of the slime on the water shield.

Emergent Plants

Aquatic plants that have most of their photosynthetic structures above the water are known as emergent plants. Therefore they are limited to the shallowest sites along the shores of our lakes. They are also vulnerable to wave action and resultant substrate movement, so most will be found along upwind or protected shorelines. One of the more common emergents is the spike rush (*Eleocharis*). This plant looks like a leafless stem originating from the sediment and terminating in a slightly swollen, pointed tip containing its inconspicuous flowers (Figure 7.6.A). This stem is produced by an underground structure called a rhizome, so the stems emerging from the substrate appear in a straight line. Since there are no large leaf blades

The Secret Life of a Lake

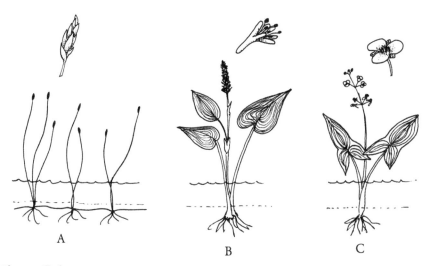

Figure 7.6
Emergents (with flower details): (A) spike rush, *Eleocharis*, (B) pickerel weed, *Pontederia*, (C) arrowhead, *Sagittaria*.

involved, it is relatively resistant to wave action and can grow in some sandy portions of the shore. In more protected shores with more organic sediments, you can find two species with upright, pointed leaves. Pickerel weed (*Pontederia cordata*) typically has a heart-shaped leaf that can vary in the amount of leaf extending in back of the petiole. All the side veins in the leaf extend parallel to the leaf margin up toward the tip (Figure 7.6.B). Pickerel weed's flower is a spike of many small attractive purple florets. The arrowhead weed (*Sagittaria* sp., Figure 7.6.C) has a similar leaf, but it usually has a pointed extension behind each side of the petiole so that the leaf looks like a giant arrowhead. Some of the side veins in the leaf are directed back toward these extensions, whereas all the side veins in pickerel weed's leaf go forward to the tip. The arrowhead's flowering spike contains relatively few (5 to 10) white flowers. When they are in flower, these two plants are very easy to differentiate.

Wetlands

Although wetlands are not actually part of our lake, they exert a major impact on the lake's water quality. In most Adirondack lakes, tributaries supplying water to the lakes pass through wetlands, and in many lakes wetlands make up significant portions of their shorelines. So it is important to understand their role in maintaining lake water quality.

Wetlands in general are defined by water exposure and a water source, as well as various soil characteristics:

> A wetland is an ecosystem that arises when inundation by water produces soils dominated by anaerobic processes and forces the biota, particularly rooted plants, to exhibit adaptations to tolerate flooding.[16]

Finer designations such as swamp, marsh, or bog, depend on whether the dominant vegetation is trees, herbs or *Sphagnum* moss, respectively. The wetlands in the tributaries and outlet around Sacandaga Lake and many Adirondack lakes are dominated by woody shrubs, such as bog myrtle (*Myrica gale*) and buttonbush (*Cephalanthus oxidentalis*), as well as royal fern (*Osmunda regalis*, Figure 7.7). There are also species of willows, dogwood, sedges and rushes, with red maple trees encroaching around the borders. Water levels are permanent and flowing in at least portions of these wetlands. Whether we call them marshes or swamps makes little difference, because the functions of the two with regard to the lake are similar.

Wetlands serve many valuable functions to surface water. They recharge and discharge ground water, buffer flow rates during floods, act as sediment traps, sequester and release nutrients, and have many valuable wildlife ben-

Figure 7.7
Wetland plants: (A) button bush—flower on the stem, fruit to the left, *Cephalanthus*, (B) sweet gale, *Myrica gale*, (C) royal fern, *Osmunda regalis*.

efits. They are also the most photosynthetically productive ecosystems on earth, even more productive than tropical rainforests.[17]

This rapid rate of biomass accumulation in wetlands is exploited by wastewater managers who create artificial wetlands to clean up effluents from storm water outfalls as well as municipal and commercial wastewater treatment facilities. Water flowing through a wetland during the growing season is stripped of its nitrogen, phosphorus and other nutrients required for plant growth, all of which are incorporated into the plants' biomass. This means that the water flowing out of the wetland will not contain these nutrients to feed algal growth in the water body receiving the outflow. However, this new biomass must be harvested and removed from the system to ensure that these nutrients will not eventually be released into the receiving water body due to decomposition of the plant material during the winter.

In natural systems, these nutrients are sequestered during the growing season, when it is important to decrease the concentrations of these limiting nutrients entering the receiving lake. However, assuming that the biomass of the wetland vegetation is not increasing, some nutrients will be released into the lake when decomposition outpaces accumulation during the photosynthetically dormant period. If the volume of the lake is flushed out rapidly, the nutrients added during the winter may be gone from the lake before the next lake growing season. Sacandaga Lake has a retention time of about 2 years, so some of the released nutrients will still be available to the algae during subsequent growing seasons.

Vascular plants play a dominant role in the littoral zone of a lake and in the wetlands surrounding a lake. If there are enough nutrients in the sediments and enough light, there will be vascular plants in a lake. In many nutrient rich lakes, they can become a nuisance when populations reach very high densities. In recent years, non-native invasive aquatic plants have become a critical problem in some Adirondack lakes, and they are discussed in Chapter 11. For the time being, Sacandaga Lake, with its long stretches of nutrient poor sand and boulder-strewn shorelines, has few stands of problematic dense aquatic plant growth. However, it has not yet been subjected to the aggressive non-native invasive weeds that have degraded so many other lakes.

8

The Consumers I: Microscopic Beasts & Immature Bugs

The producers, algae and vascular plants, fix energy from the sun by photosynthesis and make complex molecules such as sugars that fuel the metabolism of all the organisms in the lake. They are the basis of the entire food web of a lake. But which lake creatures feed on these algae and are themselves food for the trophic levels above them?

There is a vast array of lake organisms, mostly unknown to the casual observer, that are microscopic or barely visible to the naked eye. There may be hundreds, or even thousands, in one liter (about a quart) of lake water. They are wonderfully diverse, and although anatomically very simple, they interact in fascinating ways. Other larger (1 to 2 cm), animals are also present, but they have secretive aquatic stages and transient terrestrial stages in their life cycles so they escape the notice of most people.

Although inconspicuous, these organisms form a crucial link between the producers and the larger lake animals such as fish, birds and mammals that we more commonly associate with lake ecosystems. Chapter 9 will discuss these latter three groups, but now let's turn our attention to those microscopic beasts and tiny insects that are rarely seen but are vitally critical to the lake's ecology.

Protozoa

The Protozoa are microscopic, single-celled animals that are very common in lakes. They can be found freely swimming in the water column, and more frequently moving among the substrate. They ingest bacteria and other small organic particles, and have three distinct means of locomotion.

Ciliates have hundreds of small filamentous cilia on their cell surface that beat rhythmically for both movement and feeding. A common ciliate, *Paramecium*, is slipper-shaped, is covered with cilia, and has an oral groove into which larger feeding cilia direct food particles (Figure 8.1.A). Once inside, these food particles are enveloped in membrane-bound food vacuoles where digestion takes place. Some species of *Paramecium* ingest single celled algae, digest part of their cells, but retain their chloroplasts. These "kleptoplasts," as these stolen chloroplasts are called, function as they would in the alga, giving the *Paramecium* the products of photosynthesis. In exchange, the kleptoplasts receive mineral nutrients from the *Paramecium*. Once the *Paramecium* is "full" of kleptoplasts, retaining about 10 to 20, it will digest subsequently ingested algae completely, including the chloroplasts. It is amazing that such a simple organism—a single cell—has evolved such a sophisticated means of survival. For instance, how does it "know" when it has enough kleptoplasts? It challenges our definition of what it means to be a plant or an animal. Does this animal become a plant when it can gain nourishment through photosynthesis?

Another common ciliate, *Vorticella*, is funnel-shaped and has cilia only on the wide end near the mouth. However, the most fascinating thing about *Vorticella* is that it has a long "tail" with a sticky end that can attach to the substrate. When disturbed, the *Vorticella* can contract its tail into a coil like a rebounding spring, bringing it closer to the substrate. This springing action is most likely a predator avoidance mechanism, but it also increases local water turbulence that will increase food availability (Figure 8.1.B).[1]

Flagellates have one or a few long filamentous flagella that can contract and twist to move the cell about. Motion in a flagellate is jerky, whereas movement in a ciliate is a smooth glide. Perhaps the most notorious flagellate in the Adirondacks is *Giardia lamblia*, the causative agent of "beaver fever" or "backpacker's diarrhea" (Figure 8.1.D). It inhabits the digestive tract of a wide variety of warm-blooded species, and fecal contamination of a drinking water source is the major avenue of infection. It causes inflammation of the digestive tract in humans resulting in severe diarrhea, and

The Secret Life of a Lake

it is the reason one should never drink untreated water from Adirondack streams. The probability of infection is small, but if you have ever known anyone who has suffered from this malady, you know that the consequences are great. Fortunately, it does not seem to be a problem in Sacandaga and Lake Pleasant.

Pseudopodia, or long extensions of the cell membrane, are found in the common amoeba. The amoeba's elastic cell membrane extends outward, and the cytoplasm inside the cell flows into it. When a cell comes in contact with a food particle, the pseudopodia surround it and eventually engulf it into a food vacuole where it is digested. At any given time, pseudopodia can extend in many directions, but there is generally a "consensus" direction, as the pseudopodia in the trailing end contract. An amoeboid protozoan

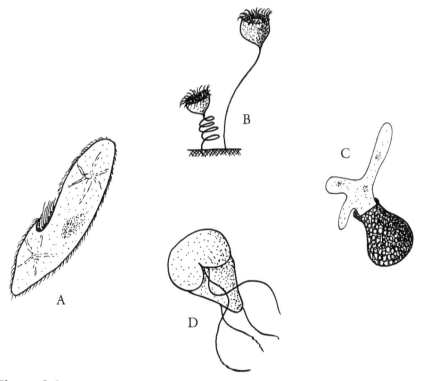

Figure 8.1
Freshwater protozoa: (A) *Paramecium*, a ciliate, (B) *Vorticella*, a ciliate, (C) *Difflugia*, an amoeboid, (D) *Giardia*, a flagellate. Most are less than 100 μm long (100 μm = one-tenth the thickness of a dime).

common in Adirondack lakes is *Diffluga*. It spends most of its time in the substrate and forms a vase-like shell around itself from small sand particles (Figure 8.1.C). During the summer, *Diffluga* forms oil droplets and gas vacuoles that reduce its density enough so that it rises from the substrate into the water column, where it spends most of the summer.[2]

Rotifers

The smallest multicellular animals in the lake are the rotifers. They range in size from about 100 microns (μm) to 1,000 μm (1,000 μm is the same as 1 mm, or about the thickness of a dime). They vary widely in shape, with some having soft bodies and others a hard shell called a lorica. Many of the loricate forms have long spines that reduce predation from larger filter feeders (Figure 8.2.B, C). In spite of this diversity in shape, they all have a ring of cilia at their anterior (front) end near the mouth called a co-

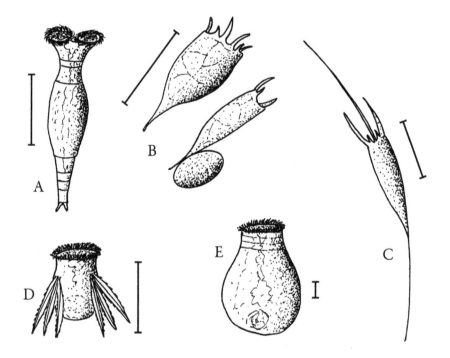

Figure 8.2
Typical rotifers: (A) *Philodina*, (B) two *Keratella*, lower one with egg, (C) *Kellicottia*, (D) *Polyarthra*—note the swimming paddles, (E) *Asplanchna*. The line next to each drawing indicates 100 μm.

rona, used for both feeding and locomotion. The cilia beat in waves, and in some common species, the corona is divided into two equal parts (Figure 8.2.A).[3] When the cilia beat in these divided coronas, they resemble rotating wheels, so early microscopists named this group Rotifera (*rota* ≈ wheel; *ferre* ≈ to bear). Rotifers can be herbivorous or predatory and are found in all freshwater habitats. Population densities of 500 to 1,000 individuals per liter are common, and in some cases can reach 5,000 individuals/L.[4]

Rotifers swim in a smooth, gliding motion, propelled by the cilia in their coronas. Their swimming paths are usually helical, and they can swim at speeds of 0.5 to 1.0 mm/sec. However one genus, *Polyarthra* (Figure 8.2.D), has 12 paddle-like structures on its sides that it can flap to produce a rapid "jump" of about 100 times its normal swimming speed. This is thought to be a predator avoidance behavior, since *Polyarthra* will jump when it senses turbulence within a distance of a few of its body lengths. During one of these jumps, it can move about 1.25 mm, which is about 10 times its body length.[5]

Rotifers have a specialized method of reproduction that allows them to expand their populations rapidly. However, to understand this exotic reproduction technique, you must first understand the usual way that most animals reproduce. When animals reproduce sexually, the males form mobile reproductive cells called sperm, and the females produce stationary eggs. The egg and sperm join to form the fertilized egg, called a ***zygote***. The genetic material in the chromosomes from both sexes joins in the zygote. To prevent a doubling of the number of chromosomes with each fertilization, a specialized form of cell division, called ***meiosis*** or reduction division, produces the eggs and sperm. This type of cell division divides the number of chromosomes in half, so that the eggs and sperm have only half the number of chromosomes (N or the ***haploid*** number) found in the parents (2N or the ***diploid*** number). For instance, humans have 46 chromosomes in each of our cells, but our eggs and sperm have only 23. When egg and sperm join, the total in the zygote equals the diploid (2N) number of 46.

Rotifers reproduce very rapidly through an asexual reproduction method called ***parthenogenesis***. Females produce diploid "eggs" through regular cell division called ***mitosis***, and each "egg" can develop directly into another female called a "daughter." (The word "egg" is used loosely here, since true eggs are always haploid.) Their "eggs" can be large, equaling one half to two thirds of their body mass in some small species.[6] In nature, rotifers typically produce an "egg" every 1 to 5 days, however a fast-growing strain of *Brachionus* has been shown to produce an "egg" every seven hours in the lab under optimal growth conditions![7] Males are much smaller than females, and are only produced under certain environmental conditions. After sensing the appropriate environmental stimulus—often poor growth conditions or a threshold photoperiod—females will produce haploid (N) eggs by means of meiosis. (These are true eggs.) If a haploid egg does not get fertilized, it develops into a "male." Males can fertilize haploid eggs, and these zygotes develop thick walls and can remain viable in this resting stage for long periods of time. Densities of these resting zygotes can be quite high in a lake bottom, with values of 100 to 4,000 zygotes per cm^2 (equal to 600 to 24,000/sq in) recorded in some lakes![8]

Rotifers are mostly filter feeders, although the large *Asplanchna* is a predator (Figure 8.2.E). Filter feeders create a feeding current with their cilia which brings small (4 to 17 μm) particulate food items such as bacteria and small algae to their mouths where they are ingested. Many rotifers can filter water equal to 1,000 times their body volume per hour.[9]

Rotifers are prey for larger filter feeders as well as predators such as *Asplanchna* and small

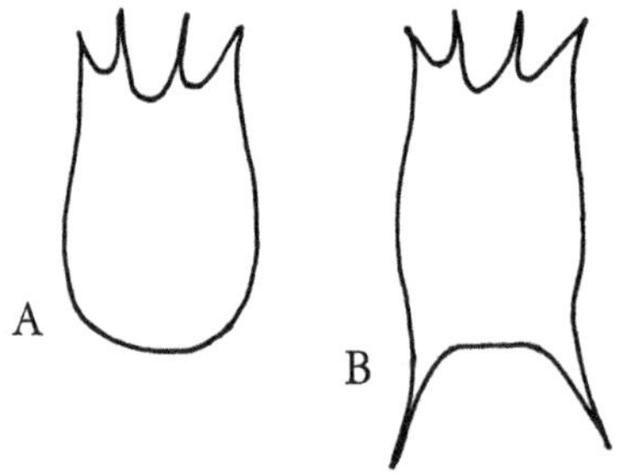

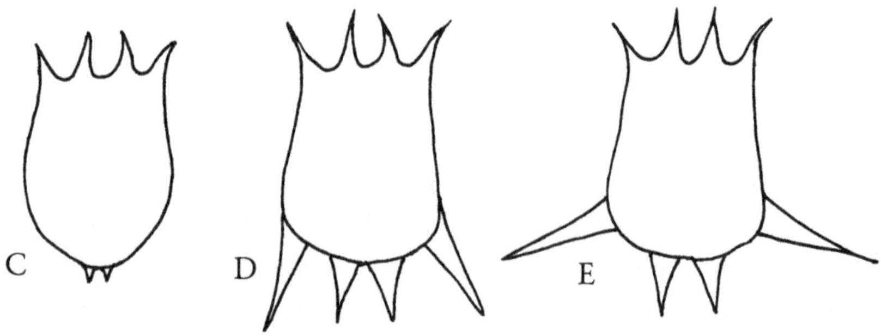

Figure 8.3
Predator-induced spines in *Keratella testudo* (A = control, B = induced) and *Brachionus* (C = control, D = induced, E = induced and extended). The loricas or shells of these rotifers are 100 to 150 μm long. (Redrawn from the original.[10, 11])

fish. Most are small and transparent, thereby reducing predation by sight, and many have spines that make predation more difficult. Some loricate rotifers have been shown to change their shape in the presence of predators. If they are grown in a medium taken from a culture of some of their common predators, many genera will produce daughters with spines that have been shown to reduce predation from both *Asplanchna* and larger filter feeders such as *Daphnia*. *Keratella testudo* will produce daughters with long posterior spines when exposed to *Asplanchna* medium,[10] and *Brachionus* will do the same (Figure 8.3.A, B).[11] However *Brachionus* takes this defensive strategy one step further. The long lateral spines are hinged, and when a *Brachionus* is touched by a predator, it will retract its corona, thereby pushing these spines out away from the body making the *Brachionus* even more difficult for the *Asplanchna* to ingest (Figure 8.3.D, E). Spines also clog the filters of the filter feeders.[12] So these simple animals, with individuals in some species consisting of fewer than a thousand cells and a brain of 15 cells, have the ability to sense the presence of a predator through some chemical cue and alter their morphology, resulting in a reduction in their risk of predation.

Various species of rotifers have different temperature ranges, dissolved oxygen preferences, and feeding niches. At any time in a stratified lake, their distributions in the lake will be segregated. In early July 2007, Sacandaga Lake was stratified with a thermocline at about 5 m (17 ft). *Trichocera* was limited to the top of the epilimnion, and *Polyarthra* was most common in the epilimnion but also found in the hypolimnion. *Kellicottia* was mainly limited to the hypolimnion, and *Keratella* was most common at the thermocline (Figure 8.4). Some rotifers are very tolerant of low oxygen levels. We have found small but stable populations in the hypolimnion of Ballston Lake, just south of the Adirondacks, when the dissolved oxygen levels were below detectable levels. Some of these rotifers even carried eggs.

Cladocerans

Cladocerans, or "water fleas," are typically larger and more complex than the rotifers, and are represented in Figure 8.5 by the genus *Daphnia*. They are also filter feeders and have a transparent clamshell-shaped covering called a ***carapace***. At the front of the carapace is a head with an obvious eye, and the bottom of the carapace is open so that food particles can enter to be filtered and ingested. They have two large appendages just behind

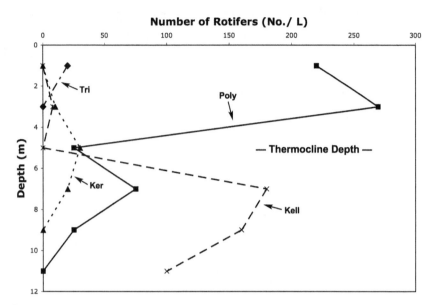

Figure 8.4
Rotifer profile in Sacandaga Lake showing *Polyarthra*, *Kellicottia*, *Keratella*, and *Trichocera*. Thermocline depth is at 5 m.

their eyes used for locomotion. When these large oar-like appendages beat, the organism jerks forward, so that cladoceran movement is always jerky. This jerky motion has given cladocerans their common name of "water fleas," although they are not insects and definitely not fleas. Inside the carapace are the filtering appendages, which contain elaborate comb-like setae that filter food particles from the water (Figure 8.5.B). When a sufficient volume of food is filtered, feeding appendages near the mouth collect it into a ball so that it can be chewed by mandibles and ingested. Whereas most of the body of the *Daphnia* is transparent, food within its gut can be quite obvious. *Daphnia* feeding on algae will have a very green gut. At the end of the body near the anus is a large post-abdominal claw (Figure 8.5.A.c). *Daphnia* can contract their abdomens so that this claw will come forward to clean out material clogging the filtering appendages. This is how algae and rotifers with large spines are "rescued" from being ingested by the *Daphnia*, since the spiny cells tend to clog the *Daphnia*'s filters more readily than spherical cells. The spiny cells and organisms are cleaned from the filters by the post-abdominal claw, are discarded and thereby survive.

The Secret Life of a Lake

Their opened carapace can create a problem for cladocerans. If an air bubble gets trapped inside the carapace, the cladoceran is buoyed to the surface where it cannot feed and it dies. One summer we had a few days of very windy weather with whitecaps on Sacandaga Lake. The next morning was dead calm, and I canoed out onto the lake. I found what appeared to be an oil slick on the water surface about 50 × 150 ft in size. After taking a sample and observing it under a microscope, I found that the "slick" was

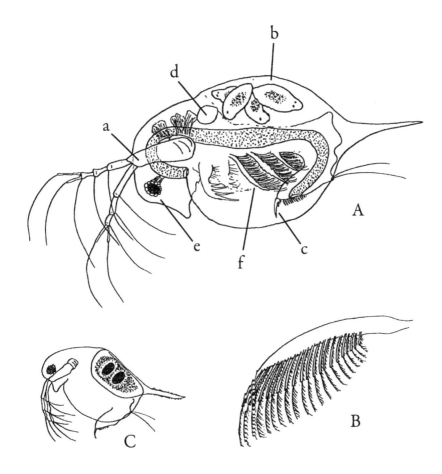

Figure 8.5
(A) Daphnia, a cladoceran (2 to 3 mm long), showing (a) swimming antenna, (b) brood chamber containing embryos, (c) post abdominal claw, (d) heart, (e) eye, and (f) filtering appendage. (B) Enlarged filtering appendage showing setae. (C) Brood chamber modified into an ephippium. (Illustrations A and C have been modified from the original.[1])

in fact a massive raft of floating dead cladocerans—millions of them—that had most likely been killed by being whipped into the air in the turbulent water. I urge all who live on a lake to venture out during calm mornings and observe the lake surface. You will be amazed at what you find.

Like rotifers, cladocerans reproduce asexually though parthenogenesis. Females produce diploid "eggs" that develop directly into adult females. Cladocerans retain these "eggs" and developing embryos in brood chambers between the dorsal carapace and the abdomen (Figure 8.5.A.b). When the embryos have developed fully, they are released from the brood chambers as smaller versions of their mother. When environmental conditions deteriorate, such as a decrease in food supply, population overcrowding or decreasing photoperiods, the adult female will produce a brood of both male and female offspring. The females then produce haploid eggs, and the males, which are typically smaller than females, will mate with the female. The resulting zygote, or fertilized egg, will form a "resting egg" (really a zygote), which is very resistant to freezing and desiccation. The resting eggs are also protected by a modification of the brood chamber. The carapace becomes thicker and encloses the "eggs" for further protection. This structure, called an *ephippium*, often includes the spine at the end of the carapace (Figure 8.5.C).

This resistant stage in the life cycle of the cladoceran is a major means of dispersal. The ephippia will float, increasing local distribution, and they are dark and easily seen by visual predators. The resistant layers of the ephippia will prevent digestion of the fertilized eggs by the birds or fish that have ingested them, so the predators expand the cladocerans' range even farther. The barbed spine at the end of the ephippia can get caught in the feathers of birds, giving them another vector of long-distance transport. They can even be dispersed by the wind. They can survive on land for years before they are immersed by a puddle or washed into a stream or lake where they can hatch and complete the cladoceran's life cycle.[13]

Cladocerans range in size from 0.2 mm, the size of a large rotifer, to the gargantuan *Leptodora*, which is almost 2 cm (¾ in) long. However, most are in the smaller part of this range (1 to 2 mm) but are still visible to the naked eye. They filter a wide range of particle sizes and are more efficient than rotifers at filtering small particles. In a lab experiment, the cladoceran *Daphnia* was able to out-compete two species of rotifers for a food source of small unicellular algae, so that eventually the rotifers were starved to extinction.[14] The filtration efficiency of the cladocerans makes them a major factor in controlling algal populations in lakes. As we will see in Chapter 10, one of the better ways to reduce algal blooms is to encourage large cladoceran populations by reducing their predation by higher trophic levels.

Some cladocerans are known to migrate vertically on a daily basis in lakes. They usually move to greater depths at dawn and return toward the surface after dark. In stratified lakes, there may be a nutritional dimension to this migration, since the hypolimnion is richer in phosphorus and nitrogen, and many prey species also occupy segregated depths. However the most likely explanation is reduced predation by visual predators. Although small, most cladocerans are large enough to be easily seen if there is enough light, and they are a prime source of food for small fish. By occupying the epilimnion only in the darkness, they reduce this type of visual predation. Since cladocerans are filter feeders, they are not constrained in their search for food by darkness. To increase their tolerance for their trip into a low-oxygen hypolimnion, some cladoceran species can tolerate dissolved oxygen levels below 1 to 2 ppm by producing hemoglobin in their blood.[15]

Like rotifers, cladocerans can sense the presence of potential predators and alter their morphology and behavior accordingly. Morphological changes occur, but are more muted than those found in rotifers. Behavioral changes include vertical migrations. In an experimental pond lacking plankton-eating fish, *Daphnia* did not show a diurnal vertical migration. However, when plankton-eating fish were introduced to the pond, the *Daphnia* quickly began to migrate vertically.[16] Subsequent research showed that the cue that initiated the daily migrations was chemical rather than visual.

Copepods

Freshwater copepods range in size from 0.5 mm to about 5 mm, and are mostly transparent. They all have a characteristically large pair of first antennae that are used for swimming (Figure 8.6). Like the cladocerans, they swim with a jerky motion. They have a single eye in the front of their large anterior segment, so early taxonomists named a common group *Cyclops* after the mythological one-eyed giant. There are two major groups found in most Adirondack lakes, as indicated in Figure 8.6.A and B.

Copepods reproduce sexually, and the males and females are approximately the same size and live about the same length of time. Under optimal conditions, the females can produce a batch of five to twenty "eggs" (really embryos) every few days. Since females can live one to several months, they can produce many hundreds of eggs during their lifetime. Eggs hatch to form a small immature stage called a nauplius (Figure 8.6.C), which develops through a series of intermediate stages into an adult in one to three weeks.[17]

Some copepods (Figure 8.6.A) ingest small particles in the range of bacteria (less than 1 μm) to macrozooplankton (greater than 1 mm). Others (Figure 8.6.B) are grasping feeders and feed on larger prey, such as algae, rotifers, larger zooplankton, small worms and even larval fish. Copepods can be quite abundant in a lake, and they have been observed in Mirror Lake, New Hampshire, to make up over half the zooplankton biomass.[18] They are selective feeders, preferring soft-bodied rotifers to spiny ones, and they are very effective predators. One small copepod, *Diaptomus pallidus*, at "average" natural densities can remove all of its preferred prey from a water volume equivalent to the entire volume of their resident pond in

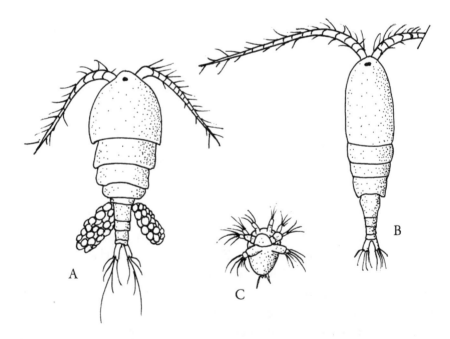

Figure 8.6
Examples of the two major groups of adult copepods in Adirondack lakes (A and B, 2 to 3 mm long). Note the egg masses behind A. An immature copepod, or nauplius (C).

one day![19] So copepods can have a major impact on species composition in their ecosystem.

Insects

Many insect groups have aquatic immature developmental stages called nymphs or larvae. Depending on the species, these aquatic stages can live from a month to a few years, whereas their terrestrial adult stage may last from a few minutes to a few weeks. The three groups we will discuss next, the mayflies, stoneflies and caddisflies are very common in both streams and lakes and are important players in nutrient cycling and the food webs of these aquatic systems. They are collectively called ***macro invertebrates***.

Because these groups are so common in most fresh water bodies, and because many of their species are very sensitive to poor water quality, they are used to assess human impacts on a stream or lake. Many species are especially sensitive to low dissolved oxygen levels as well as low levels of toxic substances, so their presence or absence will indicate the degree of stress on the sampled water body. This method of assessing human impacts on a water body is valuable, since the larvae and nymphs live in the water over a long period of time, some species over two years. If a brief episode of pollution were to occur during their life span, the populations of those species most sensitive to that pollutant would be eliminated from the stream, and subsequent sampling would note their absence. A chemical water sample describes the water quality only at the time of the sample, but it cannot assess pollutants released intermittently or even in a single spill some time in the past. The aquatic larvae will integrate the water quality over the entire period of their lifetime.

Why would some insects evolve both aquatic and terrestrial stages in their life cycles? For one thing, since the two stages occupy totally different habitats, they do not compete for resources. The aquatic habitat is more stable thermally, and the immature stages do not have to deal with the desiccating terrestrial environment. Predator pressure may also be lower in the aquatic environment. Their adult terrestrial stages are reproductive, and this environment will provide more mobility, improving their chances of finding a mate.

These groups of insects provide a major source of food for fish in both lakes and streams, and the sport of fly fishing is based on mimicking the various stages of their life cycles with artificial lures. A successful fly fisher will understand where in the steam or lake insects live and which ones are active at any given season or time of day. By matching the size, color and stage of the insect species upon which the fish are feeding ("matching the hatch"), the angler will be far more successful at catching fish.[20]

Mayflies

Mayflies (order *Ephmeroptera*) have an adult stage that fixes its wings vertically like a sail when it is at rest (Figure 8.7.A), and most species have long tails rising from the end of the abdomen. Their immature aquatic (nymphal) stage also has long tails—often three in number—as well as a single wing case and feathery gills on their abdomen. Mayfly **nymphs** from flowing streams often have flattened bodies so they can cling closely to rock surfaces and not be swept away by the current. Lake mayflies are typically burrowers and have a more rounded body cross section.

A mayfly's life cycle starts when a mated female drops her fertilized egg into a body of water. The fertilized egg may hatch immediately, or it may "rest" for a long period before hatching. This resting period is a means for the mayfly to survive conditions unsuitable for growth or survival. The fertilized egg will eventually hatch into a nymph, which will go through various growth stages that last from a number of weeks to two years depending on the species. At some environmental cue, perhaps temperature, photoperiod or some other unknown stimulus, the mature nymph will metamorphose directly into an adult. The nymph will rise to the surface by either swimming or riding an air bubble. Once on the surface, it will shed its old exoskeleton (its "husk"), and the adult will emerge with its large sail-like wings perpendicular to the water surface (Figure 8.7.A). It will take a little time for its wings to expand and dry enough to fly, but the adult will soon fly off the water surface. This transformation from the nymph to the adult is commonly referred to as a "hatch," and for some species, their hatches can be very synchronous.

Once aloft, the adult may go through one more stage during which its wings change from being grayish and fringed to clear, but eventually it flies off to find a mate. Adults of most species of mayflies cannot feed. Males form a mating swarm that can be quite large—thousands of insects or more—and females that enter the swarm are mated quickly. The mated female then deposits her fertilized eggs into the water by flying over the surface and dipping her abdomen under the surface. After she deposits her

The Secret Life of a Lake

eggs, she dies, and after the male mates, he also dies. A mayfly's adult stage is quite ephemeral, lasting from only a few minutes to a few days, which is why the group's scientific name is *Ephemeroptera*. As an example, the mayfly *Dolania americana* lives in a stream in Pennsylvania for two years as a nymph. Its hatch occurs each year in only 7 to 10 days in June, and each day it lasts only one half hour just before sunrise. During this short half hour period, nearly 3,000 flies were counted coming off one 100 m (330 ft) section of the sampled stream. The adult males of this species live a maximum of 30 min, and the females about an hour.[21]

The timing of the hatch is critical to the survival of the mayfly. Hatches are often quite synchronous, with swarms of flies appearing quickly, and then disappearing shortly afterward. Mayflies are often attracted to lights, and their massive numbers can become a nuisance to humans living nearby. In the upper reaches of the Mississippi River, mayfly swarms have been recorded so large that snowplows had to be used to clear highway bridges.[22]

Two hypotheses have been developed to explain such tight synchrony

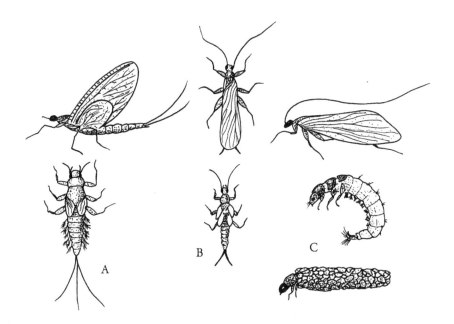

Figure 8.7
Aquatic insects: (A) mayfly adult (top) and nymph, (B) stonefly adult (top) and nymph, (C) caddisfly adult (top), larva (middle), and larva in protective case (bottom). Nymphs are 1 to 3 cm long.

in mayfly hatches. The first suggests that synchronous hatches produce larger mating swarms that increase the probability of mating. The second suggests predator satiation as the main driving force. During the hatch, resident predators eat their fill and are quickly satiated. Because the hatches are so brief, predators from nearby areas don't have time to move toward the hatch, so the wide majority of the hatch survives. Evidence showing that mayflies living in streams with predatory fish have more synchronous hatches than mayflies in streams lacking predators,[23] and the fact that parthenogenic mayflies also have synchronous hatches,[24] suggest that the satiation hypothesis is the more convincing of the two.

Mayfly nymphs are benthic organisms that may be carnivores, but are more frequently herbivores or detritivores (eating dead organic matter). Some stream species have bristles on their front appendages to filter food from the flowing water, and others spin nets. But most crawl along the bottom of the lake or stream and graze on the organic "scum"—the rich populations of algae, protozoa and rotifers, as well as organic debris—that covers the rocks and lake substrate. Population densities can be quite high. In an Oklahoma reservoir, the population of the burrowing mayfly *Hexagenia* reached densities of 150 to 960 per m^2, depending on the time of year.[25]

Sacandaga Lake does not have massive hatches of mayflies, but if you are observant, you will notice that hatches do occur. Since the mayfly nymph rising to the surface is vulnerable to visual predators like fish, hatches tend to occur under the fading light of dusk. Many evenings, if the water surface is calm, you will notice the unmistakable circles made by fish rising to the surface to eat the mayflies as they emerge from their husks and wait for their wings to dry. If you canoe out on the lake at this time, you can actually see the mayflies appear on the surface, and a short time later fly off. An angler with a fly rod and an appropriately sized dry fly will be rewarded with an exciting half hour until the hatch ceases. If you venture out the next morning while the water is still calm, you will be amazed at the number of nymph husks and dead adults floating on the surface. Certainly many more flies emerged during the night and dawn hatches. The presence of mayflies in our lakes indicates good water quality.

Stoneflies

Stoneflies (order *Plecoptera*) fold their wings flat over their body when resting (Figure 8.7.B). Their nymphs look similar to a mayfly nymph, except that they have two stout tails, two wing cases, and gills coming from their thorax rather than abdomen as seen in the mayflies. Stoneflies are mainly found in streams, but a few species can be found in lakes.

After mating, the female deposits her fertilized eggs into the water. Nymphs live from three months to three years in some large species. Upon maturity, the nymphs crawl out of the water, so there is no stage when they are especially vulnerable to predation while swimming in the water column or resting on the surface like mayflies. They molt from their exoskeleton on rocks or vegetation along the stream or lakeshore, and even on docks. Adults can live from a few days to five weeks. Mating takes place on vegetation or the ground. The mated female then returns to the water to deposit her fertilized eggs. Although this is the typical life cycle, some species emerge during cold weather, and one species completes its entire life cycle at depths of 70 m (230 ft) or more in Lake Tahoe.[26]

Nymphs are either carnivores or detritivores, eating dead leaves and the biological communities covering rocks. They seem to be more sensitive to low dissolved oxygen levels than either mayflies or caddisflies, so they are very vulnerable to polluted conditions. They are far less common than mayflies in lakes, and they are probably not an important component of the Sacandaga community when compared to other groups.

Caddisflies

At rest, adult caddisflies (order *Trichoptera*) have their wings folded above their body like a roof or tent (Figure 8.7.C). Their wings are covered with small hairs, hence the name for their order (*tricho* ≈ hair). They are weak fliers and flutter about like moths, and they have very long antennae. Their larvae look like caterpillars with a layer of armor on their heads and thoraxes. The larvae build protective cases out of sticks, leaves, bark or pebbles that can be either attached to the substrate or portable. Some species, especially lake species, form burrows. In addition to protection, the cases are also thought to aid the larva with respiration. Using abdominal movements, the larva can create a current though the case, supplying its gills with a steady stream of fresh oxygenated water.[27]

Their life cycle starts with the female laying her fertilized egg under water. The resultant larva lives a few months to a year, and after it matures, it spins a protective case and transforms from a larva into an adult in a stage known as a pupa. The pupal stage lasts about three weeks, after which the adult cuts itself free from the case and either crawls from the water or swims to the surface where it emerges from its pupal skin and flies away. Adults can live one to two months, and mating occurs on the ground or in vegetation. Mated females return to the water to lay their eggs by either skimming over the surface or diving under the water.[28]

Caddisflies are benthic feeders and in lakes usually feed on the biological community covering rocks and the substrate. As such, they are very important to the nutrient cycling within the lake. They provide food for fish, especially on their journey from the substrate to the surface after pupation, and as adults when they skim the water surface laying eggs.

Midges

Midges are another insect group common in lakes and streams. They are in the order *Diptera*, which includes mosquitoes and house flies. Their larvae live in the benthos, then move to the surface to transform into adults. The larvae of certain species of *Chironomus* are red and are referred to by the common name "bloodworms" (Figure 8.8.B). Their adult form resembles a small mosquito (Figure 8.8.A). The larva of the phantom midge, *Chaoborus* (Figure 8.8.C), spends its daylight hours in the benthos, but migrates on a daily schedule into the epilimnion at dusk where it is a ferocious predator on rotifers, cladocerans and copepods. It then returns to the hypolimnion and the benthos at dawn. Its daily vertical migration can be many meters, and it is aided by two specialized dorsal air bladders (Figure 8.8.C). This migration benefits the phantom midge by reducing visual predation by fish

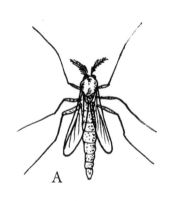

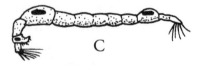

Figure 8.8
Midges: (A) *Chironomus* adult, (B) *Chironomus* larva, and (C) *Chaoborus* larva with two air bladders (black ovals). Larvae are 5 to 10 mm long.

during the daylight hours, and by allowing it access to the epilimnetic zooplankton at night. Laboratory studies have shown that these migrations are stimulated by chemical cues, since lab migrations occur in "fish-conditioned" water, but not in "fish free" water.[29] Although this larva is a grasping predator, it locates its prey though water turbulence sensed by anterior antennal hairs (Figure 8.8.C), not by vision. Therefore, it is a very successful predator in the dark.[30]

There are many other invertebrate groups in lakes such as worms, snails, freshwater clams, and others that are present in Adirondack lakes but are not as common or as important to the lakes' trophic structure as the groups listed above.

The casual observer of a lake may not even be aware of the groups mentioned in this chapter. Some are microscopic, but even the macroscopic ones usually elude all but the most curious and persistent observer. The zooplankton are very common and exert a strong influence on the algal populations, which in turn affect water clarity and therefore the aesthetic appreciation for those enjoying the lake. Although these organisms are small and simple—some are comprised of only a few hundred cells—some are sophisticated enough to be able to sense the presence of a predator and alter their shape and behavior accordingly. And there are many other subtle interactions among them that we are only beginning to understand.

The insects discussed above are not as important for what they eat, but for who eats them. They are a major source of food for fish, the lake-dwellers we are most familiar with. On calm summer evenings on your favorite lake, when fish are rising to take mayflies as they transform from nymphs into adults, canoe out and become a participant in the hatch. Look for emerging adults with their rounded wings perpendicular to the water surface. Note the amount of time it takes for the unfolded wings to dry so the adult can fly away. It can be as brief as a few seconds, or as long as a minute. If you are lucky, you will actually see a fish rise and take a fly near your boat. Know that only lakes with good water quality can provide you with such a show. We have to all do our share to keep those mayflies hatching.

9

THE CONSUMERS II: FINS, FEATHERS & FUR

COMPARED TO THE RELATIVELY INCONSPICUOUS groups mentioned in the previous chapter, this group of consumers is easily recognizable to anyone familiar with lakes. Fish have a major impact on any aquatic system they inhabit, since they not only eat algae and aquatic plants, but also all the invertebrate groups mentioned in earlier chapters. Birds and mammals are also present in most lakes, but they usually have much less of an impact. All groups discussed in this chapter are evolutionarily more advanced than the groups in the previous chapter, and all are joined by the common trait of having a backbone, or vertebral column.

Early relatives of the vertebrates contain a stiff, but flexible, supportive tube that runs the length of the body called a notochord. Just on top of the notochord they have a hollow nerve cord. An example of this group existing today is the tunicates or "sea squirts," marine animals whose adult stages look like a hollow sweet potato—hardly a relative that we vertebrates can be proud of. However, its larval stage looks a little like a tadpole, and it is in this stage where we find the notochord and nerve cord. Evolutionary advancements about 510 million years ago (MYA) led to extinct fish-like creatures with bony plates covering their bodies. Recent relatives of this group are the lampreys. The first fish with jaws appeared about 410 MYA, and the first fish with true bones appeared about 395 MYA, and they appeared to have evolved in freshwater. The bony fish are the most successful group among all the vertebrates, with about 27,000 species worldwide. Subsequent fish groups include lobe-finned fish and lungfish, which are thought to have evolved into the first group of land-dwellers, the amphibians.[1]

Fish

The fish with which we are most familiar have fins and may or may not be covered with scales. Most have a streamlined body shape to reduce drag from their viscous water environment. They are cold-blooded, meaning their body temperature is close to that of their surroundings, and they exchange oxygen with the water by means of gills.

Fins

They have both single fins—dorsal (top), anal and caudal (tail); and paired fins—pectoral and pelvic (Figure 9.1). Certain species, like trout, have a small adipose fin on the dorsal side between the dorsal and the caudal fins. The fins are supported by both spines and rays, the former being

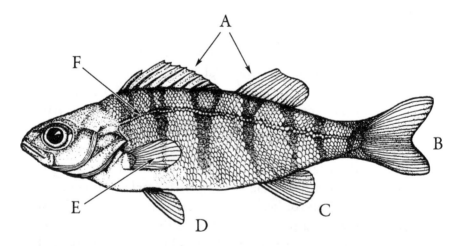

Figure 9.1
Fish anatomy of a perch: (A) dorsal fins, (B) tail or caudal fin, (C) anal fin, (D) pelvic fin (paired), (E) pectoral fin (paired), (F) lateral line.
Illustration by S. Coleman.

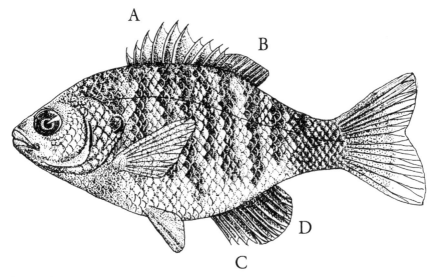

Figure 9.2
Fin spines and rays of a sunfish: Ten spines (A) and ten rays (B) on the dorsal fin, three spines (C) and nine rays (D) of the anal fin. The caudal, pelvic and pectoral fins are composed entirely of rays.
Illustration by C. DiBernardo.

rigid and sharp and the latter being cartilaginous and flexible (Figure 9.2). Spines can be found on the dorsal fins of many fish, and the anal fin of some. In rare cases, spines can occur on pectoral fins, as seen in the bullhead. Spines are mainly a defensive structure, and they may be retracted if the fish is not threatened. Anyone who has tried to grasp a bass or sunfish knows to handle the fish carefully, with your hand moving from head toward the tail to smooth down the dorsal spines so that your hand won't be punctured. Fish like trout and minnows have only rays in their fins. The dorsal and anal fins help stabilize the fish vertically, and the large caudal fin propels the fish forward. The pectoral and pelvic fins assist with steering, stopping and hovering. The numbers of spines and rays in the fins are important in distinguishing the various species of fish.

Bottom-dwelling fish use their pectoral and pelvic fins to support them as they rest on the bottom of a stream or lake. In some groups of fish, these fins developed more extensive bone structures and musculature that enable the fish to navigate better along the bottom. Eventually, the fins on some of these groups developed to the extent that they became known as "lobe-finned" fish, and it is this group that is thought to be the ancestors of the first creatures that ventured from their watery home onto the dry land.[2]

Scales

Scales cover most fish and there are two kinds. The circular cycloid scales tend to be found on smoother fish, such as trout and minnows, and the rougher ctenoid scales have jagged edges and are found on fish such as bass and perch. A magnified scale (Figure 9.3) shows many concentric rings running parallel to the outer edge. The growth of a scale is cyclic; it grows more during the summer than it does in the winter. As a result, the formation of the rings is disturbed during the winter, with the distance between

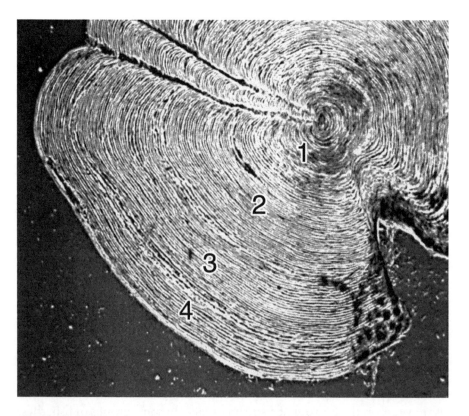

Figure 9.3
A scale of a four year old pickerel. Note the four irregularities in the scale's annuli. Each year's growth is numbered.

Table 9.1
Size and age class for fish in Lake Pleasant.

Species	Age (years)	Number of Fish Counted	Size (in)
Perch	4	1	11
	5	5	13–15
Pickerel	3	1	13
	4	2	15–18
	5	1	22
	6	2	25–27
Brown Trout	4–5	2	22
Land Locked Salmon	3–4	1	17

Note. Age was estimated by counting the number of scale annuli discontinuities.

rings decreasing and the formation of incomplete rings. If you are lucky, you can count the number of these discontinuities, and you can determine the age of the fish. Often the discontinuities are not obvious, and aging the scale is difficult. Cycloid scales are more difficult to age than ctenoid.

During an ice fishing tournament in nearby Lake Pleasant, I took some scales from some of the entries to get an idea of the age of these larger fish (Table 9.1). As a novice scale-ager, I have less confidence in my age estimates for the brown trout and land-locked salmon because of the cycloid scales in these species. Table 9.1 shows that a 13 in perch is five years old, whereas a pickerel that size is only three years old.

In most fish, there is an obvious line of slightly different scales running from the gill cover to the center of the caudal fin (Figure 9.1.F). This designates the *lateral line* of the fish, which is the location of many sensory structures that can sense water turbulence and pressure, and therefore movement in the water. Scale counts along various parts of the fish are consistent for a given fish species, no matter what size it is, because the scales grow in proportion to the size of the fish. Scale counts along the lateral line, from the lateral line to the top, and from the lateral line to the bottom are all diagnostic in determining which species of fish you are observing (Table 9.2).

Table 9.2
Scale counts for largemouth and smallmouth bass.[3]

Area Counted	Largemouth	Smallmouth
Along the Lateral Line	60–68	68–78
Above the Lateral Line	7	13
Below the Lateral Line	13	18

Feeding

The position of the mouth can give a clue as to the feeding niche of a fish. Predatory fish have large, forward-facing mouths with teeth like the pickerel (Figure 9.4.A). Bottom feeding fish like suckers have their mouths on the bottom of their head (Figure 9.4.D), and those feeding predominately from the water surface have mouths up toward the top of their heads or heads sloped back from the mouth (Figure 9.4.C). Some fish inhale their prey by increasing the size of their mouth cavity, thereby creating a water current into their mouth like a little vacuum cleaner. Such fish have forward-facing but smaller mouths, since the smaller mouth will increase the velocity of the flow into the mouth (Figure 9.4.B). To see a fish feeding like this is fascinating. Sunfish are vacuum-feeders, and you can watch them perform in an aquarium. Put a small worm on the bottom and the fish will aim its body at the worm about a centimeter or two away, and suddenly the worm disappears into the fish's mouth. A *Daphnia* swimming in the open water meets the same fate. The sunfish lines up its prey and whoosh—the *Daphnia* disappears. On the inside of the sunfish's mouth are gill rakers, cartilaginous bars near the gills that prevent the food from being expelled through the gill openings (see Figure 10.4). Young, small fish of most species eat zooplankton and use this technique to feed.

Most young fish go through an early phase when they mainly eat zooplankton. When small, they concentrate on protozoa and small rotifers. As they grow, they graduate to cladocerans and copepods, and only later will they take smaller fish and benthic meals such as mayfly nymphs, worms, or crayfish. A few adult fish, like the alewife, continue to filter zooplankton for the remainder of their lives. Larger fish select prey by their size, food value and ease of capture. They will usually choose the largest prey that they can find and capture. It takes energy for a fish to locate a prey, capture it and ingest it, and the energy required for these tasks must be more than compensated by the energy gained from the prey. *Piscivorous* fish (fish-eaters) often have a preferred prey size. For instance the large mouth bass typically prefers a prey 15 to 40% its own size.[4] Of course if they are hungry, they will be opportunistic and not so fussy. During a mayfly hatch, large fish will ingest these very small meals, but the density of the hatch and the ease with which they can locate and capture the flies makes it worthwhile for them.

Predation

However, fish are also prey and their coloration is adaptive. In most species their dorsal side is darker than their bellies. When a potential predator looks for a prey from above, the darker, upper skin surface blends in well with the dark substrate of most lakes and streams. Conversely, when a predator searches for a prey from below, the whitish belly will reduce con-

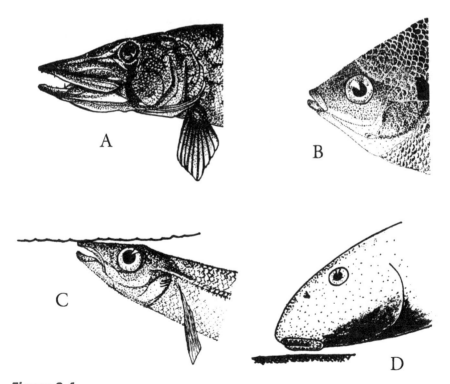

Figure 9.4
Fish mouths adapted for feeding: (A) a predatory pickerel, (B) a suction-feeding sunfish, (C) a surface-feeding silverside, (D) a bottom-feeding sucker.
Illustrations by I. Brennan, K. Maison, L. Crescenzo.

trast from the bright background illumination of the lake surface. Another advantage of this color gradient is that the light–dark coloration obscures the profile of the fish when viewed from the side, since the eye tends to favor either the light or dark pattern depending on the background. Coloration also allows some fish to better blend into their environment. Many species are "sit-and-wait" predators that hide among the vegetation waiting for prey. The vertical patterns in the perch and the mottled patterns in the pickerel blend well with dense stands of aquatic vegetation, and these species are usually found near vegetated areas.

As you might expect, the prey species have evolved mechanisms of avoidance and escape. Many zooplankton species can sense increased water currents, so that when a vacuum-feeding predator attempts to inhale them, they can "jump" out of harm's way. The paddles on the rotifer *Polyarthra* serve this function, as well as moveable spines on other rotifers and the large swimming antennae on cladocerans and copepods. Before visual predators can ingest a prey, they must first see it. The distance between the predator and its prey when the predator first notices it is called the reaction distance. This distance varies with factors such as the species of predator, the size of the prey, and illumination levels. This explains why many species of zooplankton migrate vertically toward the surface at night when it's harder to seen them, then to the darker depths of the lake during the day. It also explains why most zooplankton are almost transparent.

Even minnow-sized prey fish have predator avoidance strategies. The stickleback is so named because of the large, sharp spines on its back to discourage predation. Others have chemical alarm signals to warn of the presence of a predator. Some species have glands beneath their skin that produce a chemical signal; when released into the water, it warns other members of their species to be on the lookout for a predator. If a predator injures or devours an individual of one of these species, the alarm signal will be released into the water. Other members of that species will sense it and alter their behavior, by "freezing," schooling, or raising their dorsal fins. All these behaviors are considered defensive and will decrease their probability of being eaten.[5]

Spawning

Fish spawning typically occurs in the spring. Walleyes spawn as the ice goes out, when they move into gravelly tributaries to lay their eggs. Fertilization in all fish is external, and the fertilized eggs settle down into gravel beds where they are protected from predators. Other species make nests by flipping their tails over the substrate to clear away loose sediment and expose the underlying gravel. Nest placement is species-specific, with pumpkinseed sunfish making their nests in 6 to 12 in of water and small mouth bass at depths of 2 to 20 ft. Males then guard the nests and wait for a female to come to deposit her eggs.

Females lay anywhere from many hundreds to tens of thousands of eggs, depending on the species. More than one female may lay her eggs in a given nest, and non-nesting males will try to invade the nest to spread their sperm. Resident males try to chase these interlopers away. Along shallow lakeshores of many lakes this nest-defensive behavior can be quite obvious during nesting season. The incessant territorial battles are fascinating to watch. After fertilization, the male will remain on the nest to protect the fertilized eggs, and he may even stay to protect the larvae until they disperse. Egg predation is a major source of mortality of fish, so nest construction and defensive behavior are critical to the survival of these species.[6]

The champion nest-builder is the fallfish, the largest native minnow in the northeast. They build nests by piling stones in an elongate oval in flowing water. Nests have been described that were six feet in diameter and three feet high.[7] There are large fallfish nests in the outlet of Sacandaga Lake, though they can't compete with the size of those monster nests.

Birds

Birds are an obvious and attractive part of a lake ecosystem, but their impact is more aesthetic and spiritual than biological. During the summer, the most commonly found birds on Sacandaga and most Adirondack lakes are mallards, black ducks, mergansers, loons and ospreys.

Ducks

Mallards and black ducks are so closely related that they can interbreed. They are "dabbling ducks," which means they feed from the surface in fresh water. Although they can dive to shallow depths, they more often turn "bottoms up" to graze on the benthic vegetation in shallow water. The summer plumage of the male mallard has a characteristic iridescent green head, white collar, rust-colored breast, and a pale gray/tan body. The female is covered with mottled brown feathers (Figure 9.5.A). On the trailing edge of the wings of both the male and female is an iridescent blue patch called a *speculum* (Figure 9.5.B). Bordering the speculum on both the leading and trailing edges is a white stripe. However, both sexes of the black duck look like a female mallard, except they are slightly darker brown and there is no white border to the speculum. The mallard's tail is whitish, whereas the black duck's tail is brown. Mallards occur throughout the U.S. and southern Canada, but the black duck is limited to the eastern U.S. and Canada east of the Great Plains.[8] They both have a diet made up of about 90% plants, with the remainder as terrestrial and aquatic insects, crayfish and snails.

Both duck species breed after their first year, forming pair bonds in the winter or early spring and establishing a nest of a shallow depression in the ground surrounded by some collected vegetation. After their eggs are laid, the pair bond dissolves, and the female does the incubation and rearing of the chicks. However, there is some evidence of sustained bonding in some

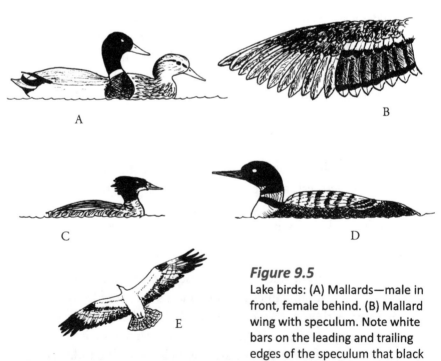

Figure 9.5
Lake birds: (A) Mallards—male in front, female behind. (B) Mallard wing with speculum. Note white bars on the leading and trailing edges of the speculum that black ducks lack. (C) A common merganser. Note the feathers protruding from the back of the head. (D) A loon. (E) A flying osprey. Note the white belly and underwing, as well as the slight bend in the wing at the "wrist."

cases. Mortality in the first and subsequent years is about 50%, and a tagging study in mallards showed a life expectancy of 1.56 years. It is presumably similar for black ducks. Mallard and black ducks can coexist, but black ducks appear to be the more hardy.[9]

Common mergansers also frequent our lakes, but they have a rust-colored head, a sharp beak and a grayish body in their summer plumage. The male has distinctive black and white plumage during its winter/spring mating season, but resembles the female during the summer and fall, and they do not stay with their brood in the summer.[10] Feathers on the back of their heads extend backward, giving the merganser's head a windswept look (Figure 9.5.C). Mergansers swim with their bodies very low in the water, and they are easily distinguishable from the two dabbling ducks. The merganser is a fish-eater and is capable of long and deep dives. Their beaks have saw-toothed edges, which allow them to catch and ingest their slippery prey. They prefer to nest in hollow trees, but can nest on the ground or in rock crevices.

A more unpleasant characteristic of ducks is their ability to act as a vector for the parasite that causes "duck itch" or "swimmer's itch," an irritating rash in humans. The life cycle of the parasite passes through both ducks and snails, so reducing the presence of either species will reduce the prevalence of the condition. Killing snails is one method of control, but a more ecologically responsible approach is to discourage the feeding of ducks. If the parasite is present in your lake, make sure that you dry your skin thoroughly with a towel as soon as you come out of the water. Allowing water droplets to dry on your skin seems to increase the risk of infection.

The Osprey

Ospreys are large raptors, larger than most hawks, but smaller than eagles. Their wingspan can reach five feet. In flight, they are easily identified by their white underbellies and whitish feathers under their wings. Their wings have a bend at the "wrists" so that the ends are angled backward (Figure 9.5.E). They eat fish, which make up about 99% of their diet. They dive into the water feet-first, with their sharp talons extended. They have been clocked at 20 to 40 mph when they hit the water, and they can catch fish as deep as three feet. They can catch and retrieve fish up to 2.2 lbs, which is remarkable when you realize that the ospreys only weigh 3.3 to 4.5 lbs! As they leave the water, they arrange the fish in their talons so that it faces forward, thereby decreasing wind resistance as the bird flies. Their success rate in catching fish was 62% in one study in California, and the parents averaged almost five fish per day to feed two nestlings.[11]

Ospreys build their nests on top of telephone poles, dead snags, or any other high, substantial site. Nests are built from sticks and can be 4 to 5 ft in diameter. The birds add to the top of the nest every year, so it can get quite large. Ospreys mate in their third year, laying 2 to 3 eggs. Young birds leave their nest after about 50 days, and mortality in the first year is about 50%. Adult mortality is much lower at 16 to 19% per year, and one banded bird lived almost 22 years!

In the 1950's and 1960's osprey populations plummeted, presumably due to the effect of DDT and its metabolites on the eggshell thickness of these top-level predators. These chemicals were bioaccumulated up the trophic levels until they reached lethal levels in our top avian predators like the eagle and osprey (see Chapter 5). After DDT was banned in 1972, osprey populations rebounded in most states, although some states still list the osprey as an endangered species.[12]

The Loon

Perhaps the bird most evocative of an Adirondack lake is the loon. It is a large aquatic bird with a black (really very dark green) head, red eye, long sharp beak and a white collar. Its breast is white and its body and wings are black with white checkering. It floats very low in the water (Figure 9.5.D). Its warbling, yodeling call is unmistakable and eerie to the uninitiated listener on a dark night. It dives for fish and, like the merganser, has barbs on its beak and tongue to grasp and eat fish under water. Because its legs are located so far back on its body, it is powerful and graceful under water but very awkward on land. As a result, its nests are built near or on the water. Its nest is a mass of vegetation at the water level on a floating island, land promontory, or even an old muskrat nest. They breed in their second year and usually lay about two eggs. Their pair bond may be sustained, with the same pair returning to the same nest year after year. Both sexes incubate the eggs.

A loon's diet is about 80% fish, with crayfish, frogs, snails and leeches making up the remainder. In certain localities, plants may make up 20% of the diet.[13] Since they feed so high on the food chain, they are also susceptible to the bioaccumulation of toxins in the environment (see Chapter 5).

The Secret Life of a Lake

They are apparently more susceptible to environmental mercury than osprey, eagles, otters and mink.[14] A more extensive discussion of this human-generated toxicant and its effect on loons is found in Chapter 11.

The loon's ecological impact on a lake is far less than its aesthetic and spiritual value. The loon is an icon of the wilderness, and it represents a hint of the primeval northern forest before the encroachment of humanity. Of course this is an illusion, since loons have adapted well to at least some human presence, and it is not uncommon to see and hear loons not far from a busy dock. But the mystical image remains and I, for one, choose to embrace it.

Cormorants

An infrequent visitor to our lakes is the cormorant, a large black piscivorous bird that looks a little like a slimmed-down loon. In some areas like the Great Lakes and Saint Lawrence Valley their populations have become so dense that they are having a major impact on the local fishery. So far we have not seen an impact on Sacandaga and nearby lakes.

Mammals

I will discuss two mammals present on Sacandaga Lake—the mink and the beaver—as examples of how such creatures have evolved into their specialized aquatic niches. Minks are semi-aquatic members of the weasel family, with an elongated body and a long furry tail. They are aggressive carnivores, equally adept on land and in the water. They eat fish, crayfish, rodents and other small prey, but they have also been known to eat muskrats and ducks. They rarely wander far from water, and they can be seen scampering along the rocky shores of many Adirondack lakes. Since minks only frequent the margins of our lakes and are very territorial, population densities remain low, and their impact on our lakes is relatively small.[15]

Beavers are the largest North American rodent, and can reach lengths of 3 to 4 ft and weights of 50 to 60-plus lbs. They have webbed feet for swimming and a broad, flat and scaly tail used for swimming, defensive warnings, and balance when they are standing upright on two feet. Their impact on their surroundings can be significant. Their huge incisor teeth can make short work of a sapling, leaving a characteristic stump with a gnawed, pointed end. They then strip the tree or branch of its bark, digesting the soft inner bark as their main food.

But beavers' main reputation is based on their ability to construct dams and lodges from branches and mud. Some of these structures are truly magnificent, with one dam in Montana measured at 2,140 ft long! Other dams have reached heights of 10 to 12 ft.[16] The dams create ponds, which provide the beaver with protection from predators. Their lodges are nearby, and the increased water level from the dams ensures underwater access to their lodges, making it difficult for most predators to enter. Their effect on our lakes is indirect. Dams on lake outlets increase lake water levels, and dams on tributaries can reduce fish spawning areas, prevent fish migrations, increase incoming water temperatures, and if the resultant pond is large enough, reduce water input to the lake through evaporation from the beaver pond.

To appreciate how well these two animals have evolved to fit into their respective niches, we should look at their skulls (Figure 9.6). Both skulls have the same bones in the same relative positions, but evolution has

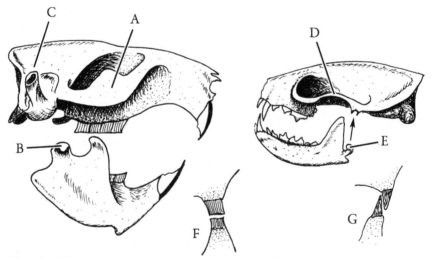

Figure 9.6
Skulls of a beaver (left) and mink (right) with disarticulated jawbones: (A, D) Zygomatic arches or cheekbones. (B, E) Jaw pivot points when articulated with the rest of the skull. Note in the mink that this joint is well defined (arrow), ensuring a consistent geometry of the teeth as the mink closes its jaw. In the beaver, this joint is loose, allowing the lower jaw much lateral motion as the molars grind the food. (C) Bony auditory channel. (F) Cross section of a beaver's jaws showing the grinding molars. (G) Cross section of a mink's jaw showing its slicing molars.

sculpted them differently to support the two different life styles. Unlike their reptilian relatives whose teeth are mostly all the same front to back, mammals typically have four different types, each modified for a different function. In the front are the incisors, usually flat and chisel-shaped for cutting. Next are the larger, pointed canines, used for puncturing and grasping. Next come the premolars, which usually have only one point or cusp, and the last are the molars with more than one cusp. Depending on the animal, some teeth may be heavily modified, or even absent.

The beaver has two huge incisors, top and bottom, that it uses to chip away wood to fell trees and gnaw away outer bark to expose the nutritious inner bark layers. As with all rodents, these teeth continue to grow throughout its life. So the busy beaver *must* gnaw frequently or its front teeth will grow so large that it cannot grind its food, and it will die. These teeth have an outer thin, orange layer of very hard enamel, which represents the cutting edge. Behind that layer is a much thicker, white layer which supports the thin enamel layer, and is composed of softer dentin. As the beaver chews, the dentin wears away before the enamel, so that the thin, sharp enamel layer is always exposed and the teeth remain sharp. Behind the incisors the canines are missing, leaving a wide gap. Using this gap, the beaver can close its lips behind its incisors so that it will not swallow water as it chews under water. The beaver's premolars and molars are flat on both the top and bottom jaw, and they are perfect for grinding food before the beaver swallows it. The joint that connects the jawbone and the upper skull is not at the same level as the teeth, but much higher on the jawbone (Figure 9.6.B). This allows all the molars and premolars to meet at the same time when the jaw closes, so that all these grinding teeth will be in close contact, making the grinding very efficient (Figure 9.6.F). In addition, this joint in the beaver is very loose, so that the lower jaw can move forward and back, right and left, further increasing the grinding motion available to the beaver.

The mink has small incisors, used to nip off parts of its prey, and large canines on top and bottom. The canines grasp the prey as it struggles, reducing the chance that it can escape. The premolars and molars are sharp and pointed, since slicing rather than grinding is more important to the carnivorous mink, and the last premolar on the top and the first molar on the bottom are enlarged and sharpened. They are offset and overlap, giving them a shearing function similar to a pair of scissors (Figure 9.6.G). The joint between the jawbone and the skull in the mink is at the same level as the surfaces of the premolars and molars so that when the jaw closes, the teeth mesh sequentially, much as scissors close as they cut (Figure 9.6.E). This joint is like a very well-defined hinge (Figure 9.6.E arrow), so that the jaw can only move up and down and not sideways, ensuring the most efficient alignment of these shearing teeth. This action provides the mink with the shearing capability necessary to cut flesh into pieces small enough to swallow.[17]

The placement of the eye sockets is also different in the two animals. The mink's eyes face forward, giving it good binocular vision necessary for pursuing and capturing prey. The beaver's eyes are on the side of its head. Since each eye can see nearly 180 degrees, the beaver can survey its entire surroundings for potential predators. This eye arrangement is typical for most predators and prey; for instance, consider the eyes of the more familiar fox (predator) and rabbit (prey).

The arch running around the eye socket attached to the front and rear parts of the skull is known as the zygomatic arch (Figure 9.6.A, D). In humans, this arch is known as the cheekbone. It is massive in the beaver and fragile in the mink. The masseter muscle attaches to the inside of the cheekbone and runs down to the outside of the large flange at the back of the jawbone. This muscle directs most of its force at the molars, where the beaver grinds its woody meal before swallowing it. Therefore the beaver requires a massive cheekbone to anchor this powerful masseter muscle. (You can feel your own masseter muscle by touching your cheek below the cheekbone and clenching your teeth.)

The mink doesn't grind its fleshy meal, but instead uses its incisors, canines and premolars to snip, pierce and slice its meal, so its cheekbone is much smaller. The temporalis muscle, which runs from the inner jawbone to the temple of the skull, not the cheekbone, exerts most of its pressure on the front of the jaw. This muscle is much more important to the mink than the masseter, because the mink must be able to capture, hold and kill its prey with its canines.[18] (You can feel your own temporalis muscle by touching your temple as you clench your teeth.)

The beaver's skull is flattened at the top, and its ears, eyes and nose are all concentrated toward the top of the skull. If you have ever seen a beaver swim, you will understand why, since it swims with just the top of its head out of the water with its eyes, nose and ears just above the surface. Note the bony channel rising from the rear of the beaver's skull (Figure 9.6.C). This auditory channel aides the beaver in hearing as it swims along the surface. It is absent in the mink. In addition, the mink's eyes are much lower in its skull than the beaver's. The mink is a terrestrial animal that can hunt in the water, whereas the beaver spends most of its active life in the water and only ventures onto land to collect branches and feed.

Although not as common on Sacandaga as beavers and minks, muskrats and otters can also be found on many Adirondack lakes. Their skulls show the same adaptations as those found in the beaver (a rodent like the muskrat) and mink (a fish-eating carnivore like the otter).

The larger consumers of fish, birds and mammals can play a major role in the ecology of a lake. The fish are by far the most important due to their greater numbers and diversity. Birds can be regionally important if their numbers are large enough, as is the case with ducks and cormorants. Since the vertebrate consumers are often at the top of the lake food chains, they are more severely affected by environmental toxicants such as mercury and PCB's through bioaccumulation. Since we humans are drawn to these animals and know so much more about them than their more inconspicuous neighbors, changes in their populations are more noticeable to us, and they can act as "canaries" to warn us of possible environmental threats. Declining populations of birds, especially raptors, brought attention to the threat of DDT and stimulated research leading to the description of the mechanism of bioaccumulation. Disappearing fish in the lakes of the Adirondacks served notice that acid rain was a widespread environmental problem. By closely monitoring our lakes and their inhabitants, perhaps we can identify potential threats earlier, when there is still time to enact effective mitigating policies.

10

COMMUNITY INTERACTIONS

In a given liter (about a quart) of lake water, there may be thousands of organisms representing perhaps 100-plus different species. For instance, Figure 6.1 shows the variety of algae found in one sample in one lake. Add to those the protozoa, rotifers, and other organisms mentioned previously and the variety is truly astonishing. Why are there so many different creatures, and how do they coexist in such a small volume? How do they interact, and what controls the diversity and population size of each species? To answer these questions, we'll revisit some of the concepts discussed earlier, such as limiting factors, a niche and the trophic structure of the community.

You'll recall that a limiting factor is anything, physical, chemical or biological that limits the survival and distribution of a species (Chapter 4). It could be temperature, water, nesting sites, predators, or any number of other factors in an ecosystem. At the nutrient level, it is that compound in the environment that is at the lowest level, relative to the needs of the organism. For instance, in Chapter 4 we showed that phosphorus levels limited algal growth during mid-summer in Sacandaga Lake.

The niche of a species is its functional role within a particular system's habitat (Chapter 5); it includes those tolerance ranges to all the limiting factors it is exposed to in a given area. For instance, a species may have a very wide temperature tolerance, but be constrained to a small portion of that temperature range due to restrictions of other limiting factors.

A major component of a niche is the way a species gains nourishment, called by some a trophic niche. This has given rise to the concept of the ***competitive exclusion principle***, which states that if two species compete for the same resource in the same habitat, only one would survive over time. In laboratory experiments, two species of the protozoan *Paramecium* can be grown in a medium with bacteria as a food source. When grown alone, both species multiply and flourish. However, when the two species are grown in the same flask, only one will survive. One species always wins, and the researchers interpreted the data to mean that the winner is a more efficient feeder on the common food source and was able to out-compete the second species over time. These experiments were carried out over very short periods of time, so there was not enough time for genetic selection to occur, which might have reduced competition.

Species Competition in Nature

When translating these simple lab experiments to a complex field situation, we do see that species have evolved to reduce competition with each other. When Darwin observed the finches on the Galapagos Islands, he noted the wide variety of bill shapes. There were many species of ground finches that ate seeds ranging in size from the small seeds of grasses to the large, hard seeds of shrubs. Birds with broad, heavy beaks, almost like those of parrots, ate the large, hard seeds, and birds with narrower beaks ate the grass seeds. Later observers found that two different species of finches growing on separate islands had beaks of similar shape, but when these two species coexisted on the same island, their beak shapes were different.[1] What might be the mechanism that would alter the beak shapes of these two species when they coexist?

Characteristics related to food gathering such as filtering appendages in filter feeders, bill shape in birds, and foraging behaviors are genetically determined, and in natural populations there is always some variability in these traits. A good indicator of bill shape is its depth, or the dimension from top to bottom at its origin at the skull. A large bill depth indicates a bill like a parrot, and a small depth a bill like a robin.

Figure 10.1 shows the bill depth distribution of two species of Galapa-

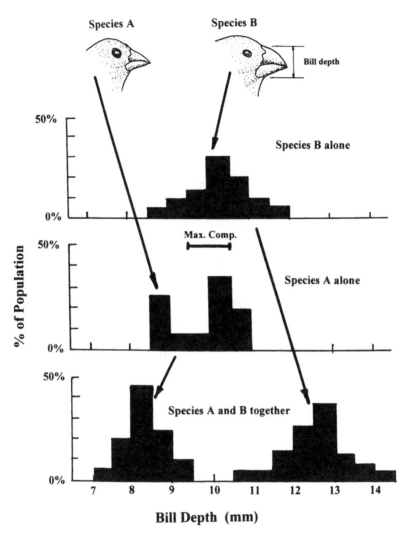

Figure 10.1
Bill depths of two species of Galapagos Island finches. The top two graphs show the distributions of bill depths of species A and B when they are found on separate islands. When both species are found on the same island, character displacement makes the bill depths of the two species diverge over time, as shown in the bottom graph. Bill depths with the greatest competition between the two species for seed sources are indicated by the short horizontal line between the top and middle graphs. Birds with bill depths in this region will get less food, and their survival rates will decline. Over many generations, species A's bill will become less deep, and species B's bill will become deeper, reducing competition for food. (This illustration has been modified from the original.[1, 4])

gos ground finches. When these two species are found on separate islands, species A has a slightly narrower bill on average than species B, but there is considerable overlap. When these two species are found on the same island, their bill depth is different—A's is smaller and B's is larger—and there is no overlap. What has happened is that when the two species first come into contact on the same island, individuals from both species with the same bill depth will compete for the same seed food resource, as indicated by the horizontal line between the first two graphs. As a result, these competing individuals will have less food available and less success in reproduction. Individuals with beak depths different from the range of overlap will have relatively more food available and should reproduce more readily. Over many generations, the average bill depths of the two species will diverge: Species A will develop narrower bills and species B will develop broader bills. Competition will be reduced, and the two species will be able to coexist by exploiting different food sources.

Given enough time, competing species can alter their genetic makeup and resultant morphology through natural selection so that eventually they can coexist. This phenomenon is called ***character displacement***. A more recent study of Galapagos finches followed the natural introduction of a species of ground finch into an island occupied by a resident ground finch species. Within just 22 years, significant divergence in bill depth occurred between the two species, with most of the change occurring after a two-year drought when seeds were scarce and competition between finches with overlapping bill depths was especially fierce.[2] Character displacement helps explain the tremendous diversity of species in many natural habitats such as our lakes.

There are other mechanisms that allow many species to coexist in the same habitat. A classic field experiment in a northeastern conifer forest studied the niches of five species of warblers, all of which lived in homogeneous spruce forests.[3] Since the spruce trees were very similar in size and structure, how could so many species of warblers, all of similar size and beak morphology, coexist if the competitive exclusion principle were valid? Researchers studied feeding habits, flight activity, and nesting sites, among other things, and found that the five species exploited different parts of the trees, or the same part at different times. For example, one species fed only on the tips of the branches at the top of the trees, another fed on the lower interior branches, and a third only at an intermediate height on both

The Secret Life of a Lake

interior branches and branch tips. Similarly, nesting sites were spatially segregated for most of the species. Therefore, by allocating these limiting resources differentially, all five species of warblers could coexist in this apparently homogeneous forest canopy. In this case, it was the birds' different behavior rather than their morphology that allowed the coexistence. Through such *resource partitioning*, as well as morphological character displacement, many species can coexist in a given habitat, allowing the large species diversity we find in natural systems.

The same is true in lakes. To the uninitiated, lakes appear homogeneous, but we know that factors such as gradients in light, dissolved oxygen and temperature create quite different habitats. For instance, Figure 8.4 shows how the rotifers in Sacandaga Lake have divided up their resource space, with *Polyarthra* primarily in the epilimnion and *Kellicottia* mainly in the hypolimnion. In Collins Lake in the Mohawk Valley of New York, we found three species of *Daphnia*, two large and one small, in a sampled depth profile. The two large species presumably filtered similar sized prey. We found one large species and the small species in the epilimnion, and the second large species in the hypolimnion, so these three species were not competing for the same food resource.

However, we do find many species of plankton, both phytoplankton and zooplankton, that coexist at the same depth. As evidence for this, recall the high algal diversity found in Figure 6.1. So the question remains as to how this many species can be consistent with the competitive exclusion principle. This was such a vexing ecological problem that the famous ecologist, G. E. Hutchinson, coined the term "paradox of the plankton" and wrote a seminal paper on the issue.[5] He pointed out that for the competitive exclusion principle to function, the environment must be uniform for a sufficient period of time for evolution or extinction to occur. Conditions in a natural lake are constantly changing—temperature, nutrients, predation rates—so that an equilibrium can rarely be reached. In addition, because there are so many different factors in a natural lake that limit populations, a successful predatory species may be limited by factors other than its food source. Yet the exclusion principle is still useful in explaining the wonderful variety of organisms in a given habitat through character displacement and differential behavior patterns.

A brief word is relevant here about invasive species, a major problem within both terrestrial and aquatic ecosystems. As the world becomes smaller with global travel and commerce, unwanted hitchhikers travel from one ecosystem, state, country or continent to another. Although many invasive species do not survive in their new habitat or survive only as a minor part of our flora and fauna, some multiply rapidly, out-competing our native species so that they dominate entire ecosystems. This occurs because their rapid population growth does not allow time for either character displacement or resource partitioning to occur with our native species. The newcomer has no resident pathogens or few successful competitors, so it expands unchecked, introducing havoc to the delicate equilibrium of the formerly balanced ecosystem.

Food Webs

Now let's consider what limits the population size of the species that are found in a given ecosystem. To do so, we must look at the community as a whole, especially the food web. We can understand niches of the creatures in our lakes by placing them in an approximate order relative to their trophic status (as in the trophic pyramid of Figure 5.2).

Figure 10.2 shows a simplified open-water food web of Sacandaga Lake. At the top are the high level carnivores—trout, bass, pickerel, walleyes—that eat other fish, and therefore called ***piscivores***. Below them are the smaller fish species—minnows, perch, smelt, bluegills—as well as juveniles of the predatory species. The smaller fish eat zooplankton and macro invertebrates, such as caddisflies and mayflies. Note that many carnivores feed at a variety of trophic levels. Larger zooplankton eat smaller zooplankton including protozoans. Most zooplankton eat phytoplankton, and the phytoplankton require chemical nutrients such as phosphorus, nitrogen and silica. This scheme is grossly simplified, and not all species or their interactions are represented. For instance, the osprey picks off an occasional large carnivore now and then. However, it does give a rough idea of their relationships, and it will be the basis for much of the subsequent discussion in this chapter dealing with factors controlling the population size of various species.

There is much discussion in the ecological literature as to whether the species in this food web are controlled from the ***top-down***, with higher trophic levels being the main factor controlling the populations of the lower levels, or from the ***bottom-up***, with nutrient availability or lower trophic levels being the most important limiting factor. This is more than an aca-

The Secret Life of a Lake

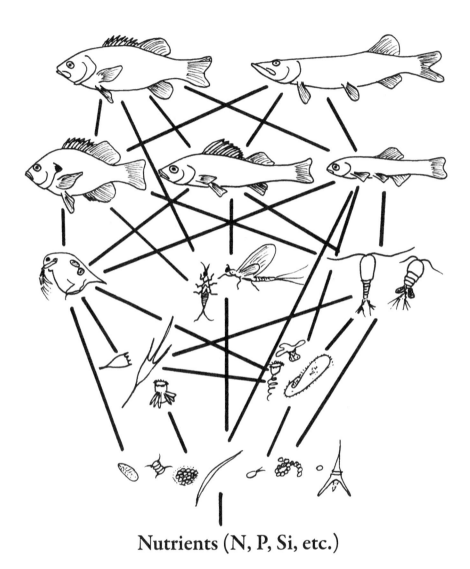

Nutrients (N, P, Si, etc.)

Figure 10.2
A simplified open-water food web of Sacandaga Lake. At the top are two piscivorous (fish-eating) fish—bass and pickerel. Below them are planktivorous fish (bluegills, perch and minnows), and below them are zooplankton and insect larvae and adults. At the base are the phytoplankton. Note that a given organism may feed at more than one trophic level. Not included are the bacteria, which provide food for the protozoa and rotifers and decompose dead organisms from all trophic levels. The nutrients at the base nourish the phytoplankton.

demic question, since recreational and aesthetic values of a lake are related to the relative abundance of its specific trophic levels. Those of us who fish are typically interested in the higher levels. But understanding how organisms at these levels are related to those at lower levels is also important. For instance, algal blooms affect both the aesthetic (water clarity, algal slicks) and recreational values of a lake, since dissolved oxygen depletion from decomposing algae can cause fish kills. So appreciating the relationships implied in Figure 10.2 can help us manage a lake to its desired state.

Bottom-Up Control: The Case of Lake Washington

One of the classic examples of successful lake management occurred in Lake Washington near Seattle.[6] During the 1940's and 1950's, the water quality of Lake Washington deteriorated, with water clarity as measured with a Secchi disc of only one meter (about 3 ft). Algal blooms were frequent, and blue-green algae were the dominant group. Fortunately for the lake, the University of Washington (UW) shared the shoreline with the city and surrounding towns, and it has one of the best academic programs in *limnology* (the study of inland waters) in the nation.

Researchers at UW had been studying the lake for decades and they understood the lake ecosystem very well. They determined that the problem came from the many surrounding towns that had placed effluent outfalls of their wastewater treatment facilities in the lake. The UW researchers showed the following:

1. These effluents were very high in phosphorus.
2. Of the total phosphorus budget in the lake, a large proportion came from these effluents.
3. Phosphorus was the main limiting factor of algal growth in the lake.

With these data in hand, the UW scientists were able to convince the surrounding towns to undertake a very expensive wastewater diversion project that bypassed the lake and deposited the effluents directly into nearby Puget Sound. The Sound is a much larger body of water with tidal currents to help disperse the effluent to minimize its ecological effects. The diversion project began in 1963 and was complete in 1968, when all wastewater

The Secret Life of a Lake

deposition into the lake was stopped.

Table 10.1 compares various water quality values of 1962 to 1966, before the diversion, with those of 1975 to 1979, after the lake had recovered. Chlorophyll levels (that indicate algal population size) were reduced to one-tenth the levels found before the diversion, and phosphorus levels were reduced by three quarters. Secchi disc water transparency increased sevenfold, and blue-green algae were reduced to a minor role among the phytoplankton. Lake Washington is a good example of a bottom-up control of the phytoplankton trophic level.

Only seven years after the wastewater diversion project was completed, Lake Washington's water quality had recovered to levels found before the significant urban development around its shoreline. Cultural eutrophication, such as that found in Lake Washington, is rarely reversed so quickly and so effectively, and certain characteristics of the lake contributed to this success. The origin of the nutrients was limited to a few easily defined sources, called *point sources*, and political pressure could be brought to bear for them to alter their practices. In addition, these sources contributed most of the phosphorus found in the lake, so there was very little internal loading from anoxic organic sediments found in many other lakes. Another factor was the relatively rapid replacement time for the water in the lake. Surface runoff entering the lake replaces 43% of the lake volume every year, so on average, the entire lake volume is replaced every 2 to 3 years. So once the input of the phosphorus stopped, the old, phosphorus-laden water was flushed out, replaced by the newer, cleaner water.

The Lake Washington case is instructive to those of us concerned about the water quality of our Adirondack lakes. We should be sampling our lakes, gathering long-term data on their water quality. If we see deterioration over

Table 10.1
Five year water quality averages, before and after wastewater diversion from Lake Washington to Puget Sound.[6]

	Before (1962–66)	After (1975–79)
Secchi Depth (m)	1.0	6.9
Chlorophyll (ppb)	31.2	3.1
Phosphorus (ppb)	64.9	16.9
Percent Blue-Green Algae	± 90	± 10

time in water clarity, the first suspect should always be nutrients. The source should be located, and if you are lucky it will be a point source that can be reduced or diverted. Otherwise non-point sources such as local urban development or agriculture within the watershed are responsible, and they are much more difficult to overcome.

Seasonal Changes

Over the course of a year in a temperate lake, there is a fairly consistent sequence of organisms present in the lake. Under the ice, the lake is relatively inactive, but it is not dormant. Light will penetrate the ice, and its intensity will depend on the thickness of the ice and the depth of snow cover. The ice reduces turbulence, so most of the algae will be small, motile forms. In some shallow eutrophic lakes with clear ice, algal blooms can even occur. In Sacandaga Lake, motile algae like *Ceratium*, *Peridinium* and *Dinobryon* can be found just under the ice. The common rotifers *Polyarthra*, *Kellicottia* and *Keratella* are also there, as are the cladocerans *Daphnia* and *Bosmina*. As you might suspect, the population size of these organisms is much smaller than we find during the summer. Fish are also active throughout the winter, as any ice fishers can tell you.

After the ice melts and the water mixes during spring turnover, the water is cold and relatively high in nutrients. Nitrate levels are very high (see Figure 4.5), and diatoms can most likely take advantage of this (as their marine relatives have been shown to do) to enhance their photosynthesis; they also can tolerate cold water temperatures well. So in most temperate lakes, diatoms are very common early in the year. Since their silicate cell walls are heavy, they tend to sink more rapidly than other algal groups, so they strip nitrates and silica from the epilimnion and deposit them in the hypolimnion. But all algae cells carry nutrients with them when they settle out of the epilimnion. So during the summer, nutrients are constantly being stripped from the epilimnion at differential rates, depending on the dominant algal groups at the time. Diatoms are favored at high silica/phosphorus (Si/P) ratios, green algae at higher nitrogen/phosphorus (N/P) ratios, and blue-greens at higher P/N ratios.[7] So a typical sequence in the Adirondacks would be diatoms early, followed by greens and other groups, and then blue-greens at the time of fall turnover when the nutrients have begun to mix from the hypolimnion and the water temperature is still warm. Although these groups are most common, all algal groups are present in the phytoplankton at lower densities throughout the year.

In the mostly meso- and oligitrophic lakes of the Adirondacks, algal biomass, as indicated by Secchi transparency, is higher in the spring and decreases throughout the growing season (see Figure 3.9 for Sacandaga Lake). The assumption is that the algae are limited by nutrients, as the settling algae over the year strip nutrients from the epilimnion into the hypolimnion. This would be a typical bottom-up control of the phytoplankton trophic level.

Top-Down Control

However, in more eutrophic lakes where there are higher nutrient levels, and even in some oligotrophic lakes, there is often a brief increase in water transparency (decrease in algal biomass) in the late spring or early summer, known as the *clear lake phase*. In Collins Lake, a eutrophic lake in the Mohawk River Valley, this phase is obvious, starting in May and lasting through most of June (Figure 10.3).

This phenomenon is now well understood to be the result of zooplankton grazing on the phytoplankton. After the spring turnover when the water warms somewhat, algae take advantage of the high nutrient levels and their biomass increases. After a period of time, the zooplankton populations respond to this bountiful food resource and rapidly expand their population size. Grazing rates increase, so that at times the community of herbivorous zooplankton can filter the *entire volume* of the epilimnion *every day*! Eventually those algae preferred by the zooplankton are replaced by species unpalatable to the zooplankton, and the zooplankton populations decrease. As a result, the algae increase once again to reduce the lake's transparency.[8] In this case, control of the phytoplankton comes from a higher trophic level, making the clear lake phase a top-down phenomenon.

Grazing pressure not only affects the phytoplankton, but other trophic levels as well. While studying many small lakes in Connecticut, researchers found that some lakes contained many species of large zooplankton—large *Daphnia* and large copepods—whereas other lakes contained only smaller species of zooplankton.[9] The common factor among the lakes with the smaller zooplankton was the presence of a small plankton-eating (planktivorous) fish called the alewife. Alewifes were not present in the lakes with larger zooplankton. Alewifes in freshwater average only about 5 in (12 cm). They have cartilaginous projections on their gill arches that act as a rake and

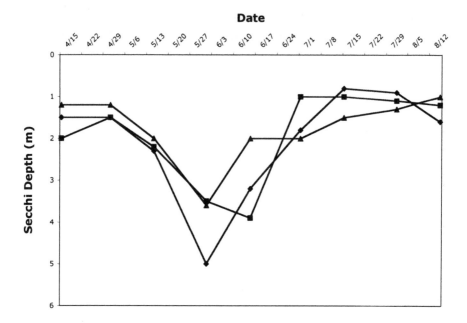

Figure 10.3
Water transparency in eutrophic Collins Lake over the summers of 1977–79. Note that the lake surface is at the top of the graph, and that deeper Secchi depths indicate greater water clarity. The "clear lake phase" in all three years starting in mid-May is most likely caused by zooplankton grazing on the phytoplankton. By July, water transparency is reduced to levels found in April by a resurgence of phytoplankton.

are therefore called "gill rakers." As water passes through their mouth and past their gills, it is filtered through these gill rakers. The alewife's "rake" is very fine, so that particles (such as zooplankton) above a certain size are sieved out and remain in the mouth cavity and are subsequently ingested (Figure 10.4).

Therefore the alewife is selectively filtering out and eating the larger zooplankton so that only the smaller species remain in the lake. In Crystal Lake, Connecticut, researchers collected zooplankton data both before (1942) and after (1964) the early 1950's natural introduction of a close relative to the alewife, with similarly fine gill rakers. The zooplankton populations shifted from larger species to smaller species after that introduction, another example of a top-down control of a trophic level (Figure 10.4).

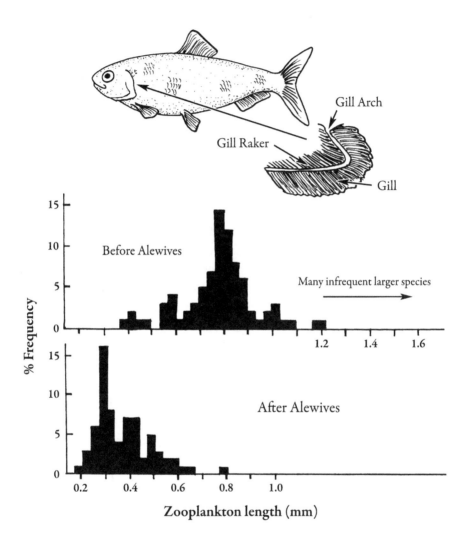

Figure 10.4
The effect on zooplankton populations by the introduction of the alewife to a lake. At the top is an illustration of an alewife, including a detail of its cartilaginous gill arch that supports its gill and gill rakers. All water passing over the gills must first pass through the gill rakers, where particles larger than the raker mesh are filtered out. The upper graph shows the size distribution of the zooplankton community found in a lake before the introduction of the alewife, and the lower graph shows this distribution five years after alewife introduction. Note that the larger species of zooplankton, those that are the most effective in controlling the phytoplankton population in a lake, are no longer present. (This illustration has been modified from the original.[9])

The Case of Lake Mendota

Perhaps one of the more ambitious attempts at controlling a targeted trophic level occurred in Lake Mendota in Madison, Wisconsin. Over the years, the lake was developing more frequent blooms of nuisance blue-green algae and resultant poor water transparency. The lake has much urban and suburban development in its watershed, and like Lake Washington, a major university with an excellent academic program in limnology on its shoreline, the University of Wisconsin. Researchers at Wisconsin had been studying the lake for years and had a good understanding of its trophic relationships.

There were no point sources of nutrients on the lake's tributaries, but instead nutrients were generated from diffuse areas of development and agriculture in the watershed. As a result, nutrient input to the lake would be much more difficult to control in Lake Mendota than it was in Lake Washington. Therefore the Wisconsin scientists were confronted with the more difficult task of trying a top-down approach to control the algal blooms. They knew that the main set of organisms controlling the nuisance algae were the larger zooplankton, especially large species of *Daphnia*. To encourage the *Daphnia* populations, they knew they had to decrease the numbers of *Daphnia*'s main predator, the planktivorous species yellow perch and cisco, a somewhat larger version of an alewife-like fish with finely divided gill rakers. To decrease the cisco and perch, they had to increase predation on them, which meant increasing the populations of the fish-eating (piscivorous) carnivores, which included walleyes, bass and northern pike.[10]

They set out on a multi-year project to manipulate the trophic structure of the lake to reduce the algal blooms. They convinced state-level fisheries to dramatically increase the stocking of walleyes and pike into the lake. In two years, the walleye population more than doubled and the pike population, which initially was much smaller than the walleye's, increased tenfold.

Then an unplanned, massive (about 90%) die-off of the ciscos occurred late in the first year of the project that was unrelated to the increased piscivore populations. Such is the case with real-life field experimentation, where not everything goes according to plan. *Daphnia* populations did increase, as did water transparency, but the improved water quality could not be directly attributed to the increased stocking. The cisco population began to rebound naturally, but then another unplanned variable entered the study.

Word got around about the increased stocking of prized gamefish, walleyes and pike, into Lake Mendota, and another top predator was added to the trophic levels of the lake—the wily Wisconsin angler. Summer fishing pressure increased from about 180,000 person-hours in 1987, the first year of the project, to 460,000 hours two years later. Walleye harvest increased from 22% of the legally vulnerable population in 1987 to 55% in 1989, an unsustainable level for the state's hatcheries to supply.[11]

Added to these uncontrolled variables was a year with atypically high nutrient runoff into the lake, making interpreting the data difficult. Nevertheless, water transparency did increase in the next few years, mainly due to a lengthened clear lake period. In summing up the results in a monograph describing the Lake Mendota project, the authors wrote, "We started this project with the idea that managing fish was the way to manage water quality. We learned that managing people has to come first."[12]

Subsequent and continuing analysis of the Lake Mendota ecosystem show that water clarity has remained variable but increased minimally since 1989. Summer water transparency is most closely related to phosphorus loading, and *Daphnia* populations to a lesser degree.[13] So both bottom-up and top-down controls are acting concurrently. These authors reiterate, almost ten years after the initiation of the project, that the problem in interpreting the effect of food web manipulations in Lake Mendota was due to the high level of angler pressure.

Otsego Lake

Another example of a top-down manipulation of trophic levels is currently being conducted on Otsego Lake near Cooperstown in central New York. Otsego Lake is long (nine miles) and narrow with an area of about 4,000 acres. Water transparency in the lake had declined due to an increase in algal populations as determined by researchers at the lakeshore Biological Field Station affiliated with the State University of New York at Oneonta. The probable explanation for the increased algae was an explosion of the population of a newly-introduced invasive exotic species of alewife, a small plankton-eating fish (see Figure 10.4).

Since this laboratory had years of baseline data on Otsego Lake on which to base their strategy, they decided to try a top-down manipulation to control the algae. In 2000, they increased the population of walleyes in the lake, and subsequent data showed a decrease in alewife density, an increase in the populations of larger filter-feeding zooplankton (the ones that eat algae), and a decrease in algae with a concurrent increase in water transparency. With the decreased rain of dead algae into the hypolimnion, oxygen levels in the hypolimnion remained higher, thereby expanding the habitat space for cold water, high oxygen-demand game fish.

So the initial stages of this intervention were quite successful. But as is the case with all field experiments in ecology, it is difficult to control all the variables. In 2007, zebra mussels were accidentally introduced to the lake, so in the future it will be difficult separate the effects of the walleye introduction from zebra mussel filtration as the cause of subsequent decreases in algal populations.[14]

These projects illustrate the difficulty of predicting the outcome of a top-down trophic level manipulation—whether initiated by knowledgeable scientists or by the accidental introduction of an invasive species—due to the complexity of these aquatic systems and the uncertainty of environmental conditions. Reasonably reliable results from trophic manipulations have been found in small lakes of a few acres where human access can be controlled.[15] The best strategy by far to confront water quality deterioration in larger lakes is through nutrient reduction (bottom-up control). But a significant effort must be spent first to identify population dynamics of the target species and its limiting nutrient, and then locating the nutrient's source and determining how to reduce it. The earlier the detection of the problem, the better the chances of success in controlling it.

Bioaccumulation, Again

I would be remiss if I didn't remind the reader of one of the most ecologically relevant aspects of the community trophic structure of an ecosystem—bioaccumulation. This process was discussed in Chapter 5, which outlined the way in which an environmental chemical can be magnified as it progresses through the trophic levels to the higher carnivores. Chemicals such as PCB's, DDT and their derivatives (Table 5.1), and mercury have been shown to cause concern both in the past and present, and we have to be constantly on the lookout for which chemical will become the toxicant of the future.

So all the inhabitants in our lakes are interacting continually with each other and their environment. Evolution has fine-tuned most interactions in fascinating ways, so that a great diversity of creatures can be found at all seasons in our lakes. There are general ecological principles that regulate the population size of most organisms, but environmental and/or human-induced perturbations can alter these systems dramatically. The more we understand the dynamics of these systems, the better chance we have of preventing unwanted water quality deterioration, or if we are clever and lucky, re-establishing a balance in an ecologically compromised lake.

11

Human Impacts

Thoreau left civilization to live at Walden Pond. More and more, however, civilization has moved into wild areas such as the Adirondacks and now encroaches on our lakes. Forests have been replaced with development. Small lakeshore camps have been replaced by year-round houses, and some by McMansions. In addition to these local pressures on our lakes, global phenomena such as air pollution and invasive species add to the mix. Have these ecological stresses affected our lakes, or has the natural resilience and stability of these complex ecosystems been able to withstand the pressures?

In this chapter we'll discuss these *anthropogenic* (human-induced) stresses. First we'll visit the more global and national pressures over which we have little control but need to understand, and then the more local issues where educated Adirondack (and other lake) advocates can have a more direct impact.

Acid Precipitation

Probably the best-known anthropogenic global pollution problem is acid rain. Combustion in power plants and internal combustion engines produces oxides of sulfur and nitrogen (SOx and NOx), which are then transformed in the atmosphere into sulfuric and nitric acids. These acids are transported in the air and are eventually deposited on terrestrial and aquatic ecosystems hundreds or even thousands of miles downwind in the form of rain, snow, and dry particulate matter.

Acidity is measured by pH units, as described in Appendix II. In Chapter 4 we saw how natural carbon dioxide in the atmosphere is dissolved in rainwater to reduce its pH to around 5.6. However, when sulfuric and nitric acids are present from air pollution, rain water can be very acidic, with pH values typically in the low 4's in the Adirondacks.[1] Some rain events in industrial areas have been recorded as low as pH 2, about the pH of vinegar, and the water in urban fogs can be even more acidic (Figure 11.1).

The fact that precipitation can be very acidic has been known for many decades, but it wasn't until the 1970's that researchers connected the dots and began to understand the effect of acid precipitation on aquatic ecosystems in the Adirondacks. They found that many Adirondack lakes contained no fish, and these lakes tended to be much more acidic than lakes containing fish.[2] A flurry of research followed, studying the origin of the acidity in lakes—some are naturally acidic—and especially the effect of soils in a lake's watershed.

Lakes with deeper soils in their watersheds tend to be less acidic than those with shallower soils, and a major determinant of acidity in a nearby lake is the type of bedrock underlying the surrounding soils. Soils derived from sedimentary rocks such as limestone, and its metamorphosed cousin marble, contain calcium carbonate, a very strong buffering agent that can neutralize the acid in precipitation before the rainwater reaches the lake. If you can recall from the discussion in Chapter 2, most rocks that make up the bedrock of the Adirondacks do not contain limestone, so most of the Adirondack lakes are very vulnerable to the effects of acid rain. Lakes in the west and southwestern parts of the Adirondacks are especially acidic, since they receive a double whammy of having very little buffering capacity in their watersheds *and* being the first recipients of the air masses coming from the Ohio Valley, a region with a high density of SOx-producing coal-fired power plants. This portion of the Adirondacks is the first area of raised topography that these acidic air masses encounter, and that is where they deposit much of their acidic precipitation.

The Effect of pH on Aquatic Communities

How does the acidity of a lake affect the creatures living in it? There are both direct and indirect effects. Data from lakes with various pH levels show that phytoplankton and zooplankton diversity and density are reduced in lakes with a lower pH. Fish species are differentially sensitive

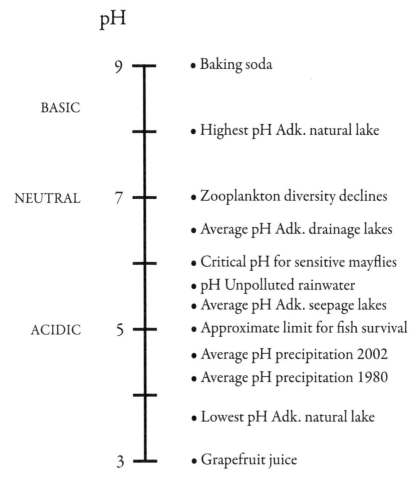

Figure 11.1
The pH scale, including various chemical and environmentally relevant reference points. Adirondack (Adk.) lake data are from the Adirondack Lakes Survey, which sampled 1,469 lakes. Drainage lakes have an outlet; seepage lakes do not. Note the thresholds for effects on the biological community, and also note that the average precipitation pH in the Adirondacks has improved between 1980 and 2002. (This illustration has been modified from the original.[1])

to low pH, with walleyes and lake trout being much more sensitive than perch and bullheads. Aquatic plants also respond differently to water pH, since some use the dissolved carbon dioxide available in more acidic water in their photosynthetic pathways, whereas others use the bicarbonate ion (HCO_3^-) found in more basic water. There are also direct effects of low pH on fish egg hatching and embryo development. Lakes with a pH of 5 or lower are typically fishless (see Figure 11.1).[3]

An indirect effect of acid precipitation involves the aluminum ion. Aluminum is a common element found in most soils, but it is mostly found in water-insoluble compounds at neutral pH's. As the acidity of the soils increases (the pH decreases) due to acid precipitation, the chemistry of the soil changes so that the insoluble aluminum compounds break down and the aluminum ion (Al^{+3}) is formed below pH 5. These aluminum ions are soluble in water, and they eventually move into surface water streams and lakes. When fish are exposed to water with low pH and high levels of the aluminum ion, water *insoluble* aluminum compounds are formed on the fish's gill surfaces, causing increased mucus production and inflammation. This deposited layer reduces oxygen exchange across the gill membrane, causing respiratory distress. In addition, certain other ion transport systems are disrupted, and the combination of these two stresses results in death of the fish.[4] When some species of fish are placed in cages in a stream or lake with low pH and high aluminum ion levels, they will die within a few days. Phytoplankton, zooplankton and other aquatic organisms are also adversely affected by aluminum ions in the water.[5]

The tributaries and lakes in the watershed of the North Branch of the Moose River, in the southwestern corner of the Adirondacks, hold waters of various pH's and levels of aluminum ions. Researchers put four species of native fish, which were at various stages in their life cycle, inside cages placed in selected streams and lakes within the watershed. Early stages of development (egg to young fish) were especially sensitive to low pH and high levels of aluminum ions, and no survival was found in three of the four test species in most of the waterways. Yearling and adult fish were also sensitive, with very low survival rates in many waters. Black-nosed dace (a small minnow) was the most sensitive species, brook trout the least sensitive, and lake trout and creek chub were intermediate.[6] It is not surprising that lake acidity has been shown to be the cause of fish loss in many Adirondack lakes.[7]

The Clean Air Act and Acid Rain
The story of acid lakes and fishless ponds is well established in the recent history of the Adirondacks, but a lesser-known sequel is now being written. The environmental movements of the 1960's nudged the government in 1970 to add a restriction for sulfur emission to the Clean Air Act, and coal-fired power plants were required to add SO_2 scrubbers to their smoke stacks. Amendments to the Act in 1990 strengthened these restrictions. Annual sulfur dioxide emissions peaked in 1973 at 28.8 million tons (mt) and have since decreased to 14.3 mt in 2003. Nitrogen oxide emissions peaked in 1983 at 25 mt and decreased to 19 mt in 2003.[8] More recent data show that SO_2 emissions have decreased further to 8.9 mt in 2007, a level even lower than the targeted 2010 emission cap of 9.5 mt required by the 1990 Clean Air Act amendments.[9] The pH of precipitation at Whiteface Mountain has increased from 4.1 around 1980 to 4.5 in 2000, with further improvement in more recent years.[10] So the Clean Air Act has had a major impact on acidic emissions in the last few decades. Have the affected aquatic ecosystems been able to respond?

Intensive sampling of many Adirondack lakes has shown a dramatic decrease in sulfate levels and more modest decrease in acidity. For instance, in Big Moose Lake in the southwestern Adirondacks, sulfate levels have decreased about 30% between 1982 and 2001. During the same time frame, pH has increased from about 5.0 to 5.6.[11] Although this increase in pH may not appear impressive, remember that the pH scale is logarithmic, and the acidification of Big Moose between 1900 and 1980 is thought to have been a decrease in pH of only about 1.0 pH unit.[12]

However, the recovery of Adirondack lakes is very uneven, with some like Big Moose showing marked improvement, but others showing little or no change. As mentioned earlier, the main variable appears to be the nature of the soil in the lake's watershed. Soils in the Adirondacks vary greatly in their depth, texture and especially the rocks from which they are derived. The rock type determines soil chemistry, which, in turn, determines the soil's ability to neutralize the effects of acid rain. This soil characteristic is called the ***acid neutralizing capacity*** (ANC), and it varies greatly among soil types. It can be depleted as the soil is continuously being exposed to acid precipitation. With the decrease in precipitation acidity due to the Clean Air Act and its amendments, the ANC in many watersheds like Big Moose is actually recovering, and the prognosis for recovery in those lakes is good, at least in the short term. However in some other lakes, the ANC shows little or no improvement, perhaps due to the amount of acidity already present in the soils of their watersheds.

So there is at least some good news in the continuing saga of acid rain in the Adirondacks. But it is important to remember that although precipitation chemistry has improved in recent years, it is still acidic. And if it continues to be acidic in the long-term, the ANC of many watersheds will eventually be depleted. Even lakes that are now showing signs of recovery may eventually become acidic again.

Sacandaga Lake is fortunate in that it contains some marble, which is derived from limestone, in its watershed. This acts as an effective buffer so that the pH of Sacandaga has remained somewhere around neutral for the last many decades. However, the calcium levels in the lake, which indicate the amount of limestone or marble in the watershed, are not especially high at about 4 ppm. This suggests that the buffering capacity in the watershed may be limited and that prolonged exposure to acid precipitation, even at the improved levels found recently, may present problems in the long-term.

Mercury

A second atmospheric pollutant is mercury (chemical symbol Hg). Some ecological aspects of Hg were discussed in Chapter 4, but its importance to Adirondack aquatic ecosystems dictates a more comprehensive coverage here. Atmospheric Hg comes from coal-fired electrical power plants, incinerators, and industrial sources. Although considered a metal, Hg is liquid at room temperature and volatilizes readily. Atmospheric Hg is found in three forms—elemental, ionic, and particulate. Ionic Hg is water soluble and particulate Hg is heavy, so these forms have a very short residence time in the atmosphere (hours to a few days), and they settle to the ground near the source. However, elemental Hg is relatively unreactive chemically, so it constitutes the largest pool of Hg in the atmosphere. Its atmospheric residence time of 6 months to 2 years ensures that this form is truly a global pollutant. In New York, only 11 to 21% of Hg deposited in the state comes from New York sources, whereas 25 to 49% comes from sources elsewhere in the U.S., and 13 to 19% comes from Asia![13] If we are to control Hg pollution in the Adirondacks, we have to act at the national and international levels. Since mercury pollution is so widespread, most aquatic systems world-wide will be affected in a way similar to lakes in the Adirondacks.

Elemental Hg is not especially toxic because it is easily excreted, but when it is deposited in wetlands or other aquatic systems, it can be transformed into methyl mercury (MeHg) by bacteria under the anoxic conditions of the wet sediments. MeHg is very toxic, causing neurological problems in higher animals and reduced reproductive capacity for most animals. MeHg is fat-soluble, so its concentration in water is very low, but algae can absorb it and concentrate it as many as five orders of magnitude, or 100,000 times! Zooplankton eat the algae, and planktivorous fish eat them, and the MeHg is bioaccumulated up the trophic levels just as DDT insecticide residues were in Table 5.1. Carnivorous fish and predatory birds at the top of the trophic pyramid contain the highest levels—two and three orders of magnitude higher than the algae, respectively—and these animals are at the greatest health risk of all the lake creatures.[14]

Because of the risk of health problems to humans as a result of Hg exposure, the U.S. Environmental Protection Agency (EPA) has set a standard for Hg levels in fish of 0.3 ppm, and they advise against eating fish with Hg levels above this standard. The Northeastern Ecosystem Research Cooperative (NERC) sampled 15,000 fish of various species in New England and New York, and found that 10 of the 13 tested species exceeded the EPA standard.[15] The highest levels were found in large predatory fish such as the lake trout and walleye, with average tissue Hg levels of 0.70 and 0.77 ppm, respectively. Northern pike, small mouth bass, pickerel, land-locked salmon and perch were also above the EPA standard, whereas brown trout were not. The New York State Department of Environmental Conservation (NYSDEC) has issued an advisory that warns all anglers not to eat more than one meal of fish per week taken from all Adirondack waters. In addition, children less that 15 years old and women who are pregnant, or might one day become pregnant, are warned not to eat any of the following fish from the Adirondacks: largemouth bass, smallmouth bass, northern pike, pickerel, walleye, and perch greater than 10 in.[16] It is ironic that some of the most prized and tasty game fish found in the Adirondacks are considered toxic by the EPA and NYSDEC.

Although all Adirondack lakes were not sampled, researchers have developed a profile of what they considered a "sensitive lake" where they would expect Hg levels in fish to be problematic.[17] These sensitive lakes have high dissolved organic carbon (such as the tan humic compounds found in Sacandaga Lake), low total phosphorus, low acid neutralizing capacity and low pH. These characteristics are typical of many, if not most, Adirondack lakes including Sacandaga.

As mentioned in Chapter 5, loons are especially vulnerable to high Hg levels since their diet is mainly fish. Because they are an icon of the wilderness, there is much interest in their well-being. Studies of loons, mainly from New England and Wisconsin, have shown adverse effects on behavior and reproduction in birds with high blood Hg levels.[18] Loons with blood Hg levels of 3 to 3.5 ppm were shown to fledge 41% fewer chicks per pair than loons with levels of 0.1 ppm. The mean blood level found in Adirondack loons was about 1.8 ppm,[19] and when this result is compared with other studies,[18] the findings suggest that the "average" Adirondack loon is reproducing at a level about 33% lower than the loons with 0.1 ppm.

To my knowledge, loons have not been sampled from Sacandaga Lake, but loons from Mason Lake and Indian Lake, both just to the north of Sacandaga Lake, had blood levels in the 1 to 3 ppm range, and loons from Piseco Lake and Ferris Lake, which are closer to Sacandaga had levels above 4![20] These levels imply real problems with reproduction. Whereas the bald eagle was the "poster child" of the fight to ban DDT from the environment in the 1970's, perhaps the beloved loon will serve the same purpose in the battle to control mercury emissions from power plants.

But shall we rend our clothes in despair about one more ecological disaster story? In fact, there are some positive subplots to this saga. Between 1990 and 2002, Hg emissions in the U.S. from both medical and municipal waste incinerators have decreased from 98 to 4.3 metric tons per year.[21] Sediment cores from lakes in Vermont show a gradual increase in Hg deposition from 1875 to about 1980 due to industrialization in the U.S., and a rapid decline since then, probably due to the better controls on particulate emissions from power plants as well as the decrease in incineration.[22] And what happens after these emission controls are imposed? Many environmental pollutants remain in ecosystems for years after deposition has stopped, but apparently this is not the case for mercury. In a locally polluted region in New Hampshire, restrictions on incinerators reduced the local annual Hg deposition from 810 μg/m^2 in 1999 to 76 μg/m^2 in 2002. Adult loon blood Hg levels dropped 64% in those same three years. In addition, both phytoplankton and zooplankton Hg levels decreased by a factor of ten in the same period.[23] The more sobering side of this story is that, while Hg from incinerators decreased from 98 to 4 metric tons per year from 1990 to 2002, Hg from coal-fired power plants has only decreased from 54 to 44 metric tons per year. This may suggest the relative political clout of large power corporations compared to municipalities and smaller industrial sources. The current national Clean Air Mercury Rule (CAMR) has a deadline of 2025 for its final goal of 70% reduction in Hg emissions. If this deadline could be accelerated, then maybe our loons will have a reprieve.

Invasive Plants

Perhaps the lake quality parameter that has caused the most concern in the Adirondacks is the spread of problematic alien invasive aquatic weeds. Among the many invasive plants that have appeared in New York, Eurasian watermilfoil (EWM, or *Myriophyllum spicatum*) and its close relatives, are currently the main threats, but water chestnut (*Trapa natans*) and curly-leaved pondweed (*Potamogeton crispus*) are also locally problematic (Figure 11.2). These three species come from Europe and Asia and have been introduced to the U.S. as ornamentals and aquarium plants. The beautiful, feathery leaves of EWM made it an attractive aquarium plant, and we assume that some people grew tired of tending to their aquaria and emptied them into a local waterway. Curly-leafed pondweed was introduced the same way. Water chestnut was an ornamental plant imported from Europe and grew in water gardens in the Boston area. It was brought to the Mohawk River Valley at Scotia, New York, about 1880, and it was released in a local lake as an ornamental because of its attractive, geometrical rosette of leaves. From there it expanded to become one of the more troublesome aquatic weeds in New York.

All three of these species have thrived in the Northeast as well as elsewhere. They reproduce profusely, pondweed and EWM through vegetative means, and water chestnut by prolific seed production. They have few natural population controls, and they are spread mainly by hitchhiking on boats and boat trailers that are towed from lake to lake. In the Adirondacks, water chestnut is currently found only in Lake Champlain. Curly-leafed pondweed is also found on the periphery of Adirondack Park, in Great Sacandaga Lake and Lake George, but it has recently been found in a few interior lakes. However, EWM is found in many Adirondack lakes, and each

Figure 11.2
Three of the major invasive aquatic plants threatening Adirondack lakes: (A) Eurasian water milfoil (*Myriophyllum spicatum*). Note the delicate feathery underwater leaves with 10 to 14 leaflets on each side of the center midrib. Leaflets terminate abruptly at tip in an approximate straight line (dashed line in leaf detail). Typically four leaves arise at each node of the stem. (B) Water chestnut (*Trapa natans*). Swollen air bladders on petioles keep the rosette of leaves floating on the lake surface. Small underwater flowers give rise to large (± 1 in) "nuts" with four large pointed and barbed spines (bottom center). Underwater "leaves" are highly dissected to increase surface area. (C) Curly-leaved pondweed (*Potamogeton crispus*). Note the wavy margins of the leaves, making them resemble a miniature lasagna noodle.

year it spreads to a few more.

When it becomes established in a lake, EWM spreads by fragmentation of stem tips, which float with the wind, then sink to the sediment to establish new populations. Fragmentation can occur spontaneously, or it can be the result of motorboat propellers or wave action. It can form dense stands that are impossible to swim through or even penetrate with a motorboat, since propellers get entangled with the stems. It will grow in organic sediment, sand and gravel, and it will grow in water up to 5 m deep (about 17 ft), depending on water transparency. In other words, once it gets into a lake, it can become a real problem to lake users. Just imagine your favorite lake with impenetrable masses of aquatic weeds growing in all areas shallower than 15 ft!

Control of Aquatic Invasive Plants

Controlling established populations is very difficult. The Adirondack Park Agency discourages the use of chemical herbicides in the park, and even if allowed, herbicide control is only temporary and very expensive, even for a small lake. In Collins Lake, a 54 acre urban lake in the Mohawk River Valley, a single treatment of the herbicide Sonar to control EWM for a few years cost about $60,000. In the larger Saratoga Lake (4,000 acres) outside the southern border of the park, a plan was developed to treat about a third of the lake for EWM with herbicide for three consecutive years at a cost of $1.4 million.[24] The cost of the herbicide alone was $700,000. Mechanical harvesting is another strategy for control, but it is also expensive and temporary. Harvesters can cost well above $150,000, and Saratoga Lake has two. Add to that the cost of labor, maintenance and fuel.[25] Mechanical harvesting is essentially like "mowing the lawn" of EWM and requires repeated effort. In a lake with a young invasion that is not completely colonized, mechanical harvesting will actually increase its spread, since harvesting creates many fragments.

We are fortunate in the Adirondacks, since the invasion of EWM is relatively recent. Most lakes are not infested, and populations in those that are have not yet expanded to their full ecological limits. If an invasion can be caught early, a program of intensive hand-pulling by divers can control the spread of EWM and even reduce the area of infestation. The goal is to hand-harvest intensely for a few years until the EWM population is small enough so that it can be controlled with a more modest program of "maintenance" pulling. This technique is also expensive, but it has the advantage of being more ecologically "friendly" to the native vegetation and less disruptive to lake use. If the invasion is caught very early, it is theoretically possible to eliminate the weed from a lake entirely.

This method was first used in the Adirondacks in Lake Colby, a 272 acre lake near the Village of Saranac Lake. Hand harvesting started in 2002, and by gradually improving their techniques and efficiency, they planned to reach the maintenance stage by 2009. Hand-pulling was supplemented by the use of mats placed on the sediment, called benthic barriers, which block the sunlight and kill the covered plants. The cost of this entire program from 2002 through 2007 was $76,000, funded mainly by state and local watershed funds.[26]

In 2004, a similar program was initiated on the much larger Upper Saranac Lake, with a 44 mile shoreline compared to Lake Colby's 3.3 mile shoreline. In an intensive three-year program of hand-pulling and benthic barriers, they were able to reduce EWM populations to a fraction of their original area and have now moved into their reduced maintenance stage. The initial three years cost about $1.6 million, however subsequent years will be less expensive, though still over $200,000 annually. Although the Upper Saranac Lake program seems prohibitively expensive, compare it with the herbicide program on the smaller Saratoga Lake (23 mi shoreline) that will provide only temporary relief and require expensive interventions in the future. By catching the infestation early, Upper Saranac Lake now has the possibility of long-term control of this nasty weed at a relatively modest cost.[27] In Lake Mascoma, a five mile long, 1,114 acre lake in New Hampshire, effective control of EWM was achieved using volunteers from among their lake association membership, because they caught the infestation early.[28]

Given the cost and difficulty of controlling EWM once it becomes established in a lake, it is far more cost effective to prevent its introduction. Education of boat owners about their responsibility to clean their boats and trailers as they move from lake to lake is critical. Many lakes have ongoing programs in which "lake stewards" inspect boats and trailers entering lakes at public boat ramps. Whereas many states have strict laws regulating the transport of invasive aquatic weeds, New York in 2010 had only one law that regulated water chestnut transport. As a result, the Town of Lake Pleasant passed its own local law prohibiting aquatic weed transport in 2009. Lake stewards in other towns are limited to educating and cajoling boat owners into cleaning their boats before they enter a water body. However, most boat owners have been cooperative, at least on the ramps in Sacandaga Lake. Larger lakes with many points of access have a much more difficult task of preventing the introduction of these weeds.

But we can't rely on the good intentions of boat owners or the constant surveillance of our lake stewards. We lake enthusiasts must educate ourselves to recognize these problem species and be on constant lookout for them. Lakeshore owners should patrol stretches of their shoreline annually, because the earlier an infestation is identified, the more likely the possibility of control or eradication.

In the Adirondacks, there is a group with the specific mission of identifying and mapping the spread of invasive plants in the Park, both aquatic

The Secret Life of a Lake

and terrestrial. The Adirondack Park Invasive Plant Program (APIPP) is funded jointly by the NYSDEC, the Nature Conservancy, and the Adirondack Park Agency, and it trains and coordinates volunteers who survey their own lakes. By 2010, 302 lakes had been surveyed, and 80 found to contain invasive aquatics. As new lakes were added to the number sampled, the number of those new lakes found to have invasive weeds has been relatively small.[29] So the invasives continue their march into the Adirondacks, though their rate of expansion is not explosive.

Purple Loosestrife

A wetland species that is creating problems in the Adirondacks is the purple loosestrife (*Lythrum salicaria*, Figure 11.3). Once it becomes established in a wetland, it expands to exclude most of the native species that are important to our native wildlife. It has a beautiful bright purple floral spike, and it is so attractive that people plant it as an ornamental in their gardens. Gardeners can purchase it from seed catalogs even though the plant is designated as a problematic invasive in most states, including New York. Some are sold as sterile hybrids, but such claims have been shown to be false. The following quote is taken from a U.S. National Park Service web page:

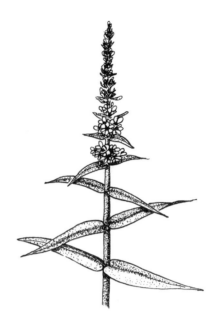

Figure 11.3
Purple loosestrife (*Lythrum salicaria*). Note the terminal spike of purple flowers and the leaves arranged opposite each other on the stem.

> "Guaranteed sterile" cultivars of purple loosestrife are actually highly fertile and able to cross freely with purple loosestrife and other *Lythrum* species.[30]

Some people consider an infested wetland beautiful, with an uninterrupted sea of purple flowers. However, that is true for about one month a year. For most of the year it is a sea of dead stalks or a sea of monotonous green shoots reminiscent of a cornfield. Compare that to the wonderful diversity of our natural wetlands, which have far greater benefit for our wildlife.

Once it becomes established, control is difficult. It has a very robust root mass that must be excavated—you can't control it by merely cutting off the top. A typical root system can produce stems for years, even though stems are cut annually before they produce seeds. A single plant can produce up to 2.5 million seeds, which are long-lived.[31]

In recent years, a biological control has been developed using insects that eat loosestrife. Insects are commercially available and control has been successful in some cases, but results are variable and the insects are expensive.[32]

Invasive Animals

Plants are not the only invasive species of concern to the Adirondacks. Exotic animals are also on the prowl. The most infamous is the zebra mussel (ZM, Figure 11.4.B) and its close relative the quagga mussel. Both arrived to the U.S. from Europe in ship ballast tanks, and first became established in the Great Lakes around 1985. Zebra mussels have since spread through the Mohawk River Valley into the Hudson, where they were first observed in 1991. By the end of 1992, ZM biomass exceeded that of all other heterotrophs in the Hudson, evidence of their incredible reproductive capacity. Their populations can reach an excess of 10,000 individuals per square meter. As a result they create major problems by fouling boats, water intakes and power station intakes—just about anything that is a solid substrate for the ZM larvae to settle on.

ZM's are filter feeders, and by measuring the filtration rate of a single mussel and multiplying that by their average density, researchers found that during a population peak, the *entire volume* of the Hudson River estuary can be filtered by ZM's *every day* during the summer! They eat phytoplankton and small zooplankton, so water transparency increases 20 to 100%. Since water transparency affects aquatic plant distribution, plants extend their range into deeper waters as well as increase their biomass.[33, 34]

Are Adirondack lakes vulnerable to ZM's? Apparently some are but many are not. Zebra mussel larvae are practically impossible to control at

The Secret Life of a Lake

boat ramps. They are microscopic and will survive in fishing boats' live wells as well as the cooling water retained in powerboat motors. They also may be present in containers of bait fish taken from infested waters. So we can assume that most Adirondack lakes will eventually be exposed to ZM's. However ZM's are mollusks that have a calcium-rich shell. Fortunately for many Adirondack lakes, ZM's aren't all that good at extracting calcium from the water, and most researchers who work with them think that the lower calcium limit for their survival is somewhere around 20 ppm.[35] Because of the Adirondack's geologic history, many lakes have calcium levels well below this threshold, including Sacandaga Lake (about 4 ppm). An interesting irony is that the low calcium levels that make our lakes more susceptible to acidification may save those same lakes from ZM infestation. This may be wishful thinking, however, since most invading species are very adaptable; so there is a possibility that ZM's may be able to eventually evolve to survive at these low calcium levels. However, many Adirondack lakes have lots of limestone and/or marble in their watersheds, and calcium levels above 20 ppm, and those lakes are vulnerable now.

Other, less conspicuous animal species are also spreading into the Adirondacks. In 2008, the spiny water flea (SWF) was first reported in the Great Sacandaga Lake, and it arrived in Sacandaga Lake (25 miles north) in 2010. This small creature—about 1 cm long (½ in)—is related to our

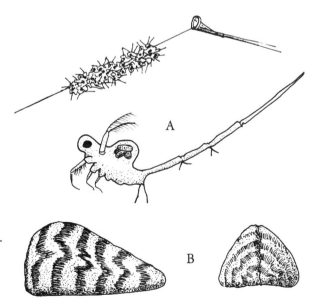

Figure 11.4
Invasive animals threatening Adirondack lakes: (A) The spiny water flea (*Bythotrephes longimanus*). Note the long terminal spine that makes its total length almost one-half inch. Dense populations can foul fishing lines (top). (B) A zebra mussel (*Dreissena polymorpha*), side and front views (1 to 1.5 in).

native water flea, *Daphnia*, but it has a very long spine at the end of its body (Figure 11.4.A). It reproduces rapidly and can soon dominate the large filter-feeder level of the food web at the expense of our native species. It eats *Daphnia* and other similar-sized zooplankton, and has caused the decline and elimination of some native zooplankton species. Its long spine makes it unpalatable to small fish or larger filter-feeders, and has caused a decrease in the number of small perch in some areas of Lake Michigan. Its long spine gets entangled with fishing lines so that it can form elongated blobs clinging to the line so dense that they can clog fishing rod guides.

The SWF is spread from lake to lake in bait buckets, live wells, anchor ropes and bilge water, so it will be very difficult to limit its expansion into other lakes.[36] However, it's not certain that this invader will be as problematic as we first thought. In its native Europe, it is an important food item for many of the same fish species we have in our own lakes. So unlike many invasives, we already have resident trophic controls to the population. In addition, in these surveyed European lakes, *Daphnia* populations are not as adversely affected as some are predicting for U.S. lakes.[37] Since the SWF appears to prefer cooler water—it was found between 15 and 25 ft deep in Sacandaga Lake—those recreational lake-users who do not troll deep in the lake may not even notice its existence. The population in Sacandaga Lake remained very small during the summer of 2011, so it has not expanded dramatically within the one year since its introduction. We'll just have to wait to see how our native ecosystems accommodate this new intruder.

Not all exotic animal species are problematic, however. Brown trout came from Germany, and rainbow trout came from the northwestern U.S. Few people who fish would consider these species a problem—unless there are too few of them in their favorite fishing hole. Both of these game fish species are stocked annually in Sacandaga Lake by NYSDEC. But these species do outcompete our native brook trout and push them into smaller tributary streams. Smallmouth and largemouth bass are also thought to have been introduced into the Adirondacks from their more southern and western native distributions.

In the Sacandaga–Lake Pleasant lake system, rainbow smelt, a 7 to 8 in bait fish, was accidentally released around 2000 and has become well-established. The smelt are thought to be responsible for the demise of walleye reproduction in those lakes.[38] In response, NYSDEC then stocked landlocked salmon (a strain of Atlantic salmon) in the lakes to prey on the smelt, and more recently they abandoned salmon stocking in favor of lake trout. Time will tell how these new introductions will pan out and what their effect will be on native and previously established species.

Nutrient Enrichment

Another potential threat to lakes caused by humans is nutrient enrichment, or cultural eutrophication. Development around a lakeshore will reduce forest cover and replace it with asphalt and lawns. Both will increase precipitation runoff into a lake, and with that come nutrients. In a forest, the soil is typically porous and contains much organic matter. Water penetrates deeply and is held in the soil by the spongy organic layers. Trees and native shrubs with deep roots prevent erosion of sediments into the lake, and these plants absorb much of the water. Contrast this with a parking lot or street, where water quickly flows along the surface, picking up any material it contacts as it makes its way to a stream and eventually a lake.

A shoreline lawn is not much better than a parking lot. Beneath the grass, the soil is often compacted, so runoff is enhanced at the expense of penetration. Grass roots are not deep, so water that gets past the shallow root system is free to move downhill toward the lake. If the lawn is fertilized, excess nutrients will eventually make their way into the lake, as well. As pointed out in Chapter 4, phosphorus is the main limiting factor for algal growth in Sacandaga Lake, as well as most other Adirondack lakes, so if lakeshore owners feel the necessity to fertilize their lawn, they should use a fertilizer that contains little or no phosphorus. A better solution is to reduce the area of lawns and provide a strip of natural, deep-rooted vegetation between the lake and the lawn to capture nutrients leached from the lawn.[39]

Septic systems are also a potential source of nutrients to a lake, so these systems must be tested frequently for leaks. Recent upgrades in regulations for the placement of septic tile fields require larger setbacks from the lake shoreline, thereby reducing the chance of contamination. The best solution is to have all lakeshore properties connected to municipal wastewater treatment facilities, but this is not possible in many cases. On the shores of Sacandaga Lake is a large campsite, Moffitt's Beach. Its approximately 260 campsites are serviced by two wastewater facilities sited within the campgrounds that are considered secondary treatment—they eliminate particulates and take out some, but not all, nutrients and bacterial food.

Since 1968, effluent from both facilities has been pumped directly into the hypolimnion of the lake after chlorination.

Human Impact on Sacandaga Lake

Has the dense human population of the hotel era in the Speculator area around 1900, along with the currently expanding lakeshore development, affected the water quality of Sacandaga and nearby lakes? The answer is that it is hard to tell. Without reliable long-term data, we are left with historical anecdotes and spotty comparisons. One indicator about which we have some long-term information is the presence of lake trout. From the historical record (Chapter 2), we know that the area was rich in game and fish around 1900. The quote from the Sturges Hotel source stated that in "a few short hours" they could catch, "sufficient lake trout to feed from 150 to 200 guests." We're not sure whether the trout were caught in Sacandaga Lake or Lake Pleasant, or whether the fish were really lake trout. However, more recent anecdotes from relatives of long time members of the community whom I have interviewed confirmed that lake trout were very common in the early 1900's. What we are certain about is that there are no longer sustainable populations of lake trout in either Sacandaga or Lake Pleasant, but they are still present in nearby undeveloped Fawn Lake.

Lake trout are native to the Adirondacks. They are considered a cold-water fish and are usually found in the summer near the bottom of well oxygenated lakes.[40] The hypolimnions of both Sacandaga and Lake Pleasant have very low dissolved oxygen (DO) levels (below 1 ppm) for at least a few weeks in September, whereas the hypolimnion of Fawn Lake retains at least 4 to 5 ppm DO all year. This could explain the current distribution of lake trout in the three lakes, but was "human impact" the main reason for the deoxygenated waters in Sacandaga and Lake Pleasant? Nutrient enrichment from human activity would increase algal growth, adding organic material to the sediment that would be decomposed by bacteria, contributing to the decrease in DO in the hypolimnion. But we have no reliable data on oxygen profiles in these lakes before the 1932 Biological Survey of the Upper Hudson Watershed.[41] In that survey, Sacandaga Lake and Lake Pleasant were sampled only once, on July 28, over a month before severe hypolimnion oxygen depletion occurs today. Secchi transparency and DO levels in both lakes in late July of 1932 were within the range of present-day values at the same time of the year. Although lake trout were listed as "common" in both lakes in the 1932 survey, we do not know whether there was natural reproduction that might have sustained some of the original population, or whether they were the result of stocking of the lakes in 1930 and 1931.

Therefore we are left with only the historical record of the presence of natural populations of lake trout, along with the assumed existence of an oxygen-rich hypolimnion around 1900, as evidence to suggest that lake water quality in these lakes has been degraded by human activities. The surviving population of lake trout in undeveloped Fawn Lake supports this assertion. But if the assumed water quality decline did happen, it most likely occurred before the 1932 survey. As stated above, the water transparency and DO profile values collected then are within the same range as current values, so a *significant* decline in water quality has apparently not occurred over the last 75 years.

Some long-term lakeshore residents have observed that there is more "scum" on submerged rocks, and that native weed beds have expanded in the decades they have been enjoying the lake. We have no data to confirm these observations, and other residents have recalled that the rocks were covered with "scum" many years ago. So we are once again confronted with conflicting anecdotes. We must also realize that lake eutrophication is a natural process, but it usually occurs at a very slow rate.

The above discussion illustrates the importance of gathering and compiling data on your favorite lake so that a decrease in water quality can be noted and documented early. This increases the probability that an intervention would be successful. Although acid rain and mercury contamination must be dealt with at the national level through legislation, nutrient issues are typically local. Appendix V lists some strategies you might use to begin assembling information on the possible human impacts on a lake. The longer the term of the data set, the more reliable it will be, since yearly variations will tend to mask trends. We cannot rely on anecdotes to convince a municipality to extend a sewer line, but reliable data may be more successful.

We are fortunate in the Adirondacks because our lakes are still in relatively pristine condition. Only with education of lake users about the main threats to our lakes along with continued vigilance can we hope to keep unwanted nutrients and invasive weeds from spoiling these jewels.

12

The Future

WHAT DOES THE FUTURE HOLD? The 1922 Nobel Physics Prize winner Niels Bohr once said, "Prediction is very difficult, especially about the future." If a Nobel Prize winner has a problem predicting the future, what are *our* chances? However, at least in the short term, we can be fairly sure that current trends will continue.

Climate Change

I have not yet discussed global climate change, which in the Adirondacks certainly means warmer and wetter weather. Although climate change is occurring as I write this and some of its effects have already been well documented, its potential for disturbance of our lakes and watersheds will become much more evident in the future.

Atmospheric Carbon Dioxide

Atmospheric carbon dioxide (CO_2) levels have risen dramatically and rapidly in the last half-century, most likely due to human activity. When I was a graduate student in the 1960's doing research on photosynthesis, we all knew that the atmospheric CO_2 concentration was about 320 ppm. In 2011, it was approaching 390 ppm, an increase of nearly 22% in my professional lifetime! Carbon dioxide is a greenhouse gas, one that absorbs the heat radiation emitted by the Earth and re-radiates part of that absorbed energy back to Earth. As the atmospheric CO_2 levels increase, this increased re-radiation will tend to warm the Earth.

As evidence of the connection between atmospheric CO_2 and the Earth's surface temperature, we can see a signature of the greenhouse effect in observed data. Of all the factors influencing the Earth's surface temperature, solar radiation is by far the most dominant. Greenhouse warming is less important, but it occurs 24 hours a day, whereas solar input is limited to the daylight hours. So the greenhouse effect will be more prevalent during the night compared with the day, and also at times during the year (winter) and regions of the Earth (polar areas) with relatively lower solar input. Observed data confirm that this is occurring. The data show that nighttime temperatures are increasing faster than daylight temperatures. Polar regions are warming more rapidly than tropical, as evidenced by melting Arctic sea ice, thawing permafrost and measurably thinning glaciers.[1] And winter temperatures are increasing more than summer temperatures. For instance, in the northeastern U.S., average temperatures have increased 1.5 °F between 1970 and 2000, with winters warming most rapidly at about 4 °F during this period.[2]

To some people these short-term temperature increases in the low single digits may not appear to be a serious problem. After all, New Yorkers know that humans have been known to survive in Georgia and Florida. However, long-term predictions are truly sobering. Not only is the concentration of atmospheric CO_2 increasing, but the annual *rate* of increase is also increasing. In the 1990's this rate was 1.5% per year, in the past decade it was 2.0% per year, and from 2009 to 2010 it was 2.3%.[3]

At this time there appears to be little political will in the U.S., or financial and logistical capabilities in the developing world, to make the massive commitment necessary to halt or even significantly slow the release of CO_2 into the atmosphere. So in the short-, mid-, and maybe even long-term, there is only a slim chance of reducing the increasing trajectory of atmospheric CO_2 levels.

Feedback Loops

In addition to the direct production of CO_2 through fossil fuel use, there are many feedback loops lurking in the background. In Chapter 2, I discussed the positive feedback that can cause glaciers to advance and retreat by altering the reflectivity of the Earth's surface to solar radiation. Similarly, the currently decreasing areas of polar ice and glaciers reflect less of the sun's energy away from the Earth, thereby heating up the surface and melting more ice.

Other, more subtle and less well understood positive feedbacks include the melting of the permafrost in near-polar regions that allows microbes to decompose the richly organic soils and peats thereby releasing more CO_2,[4] reduced CO_2 absorption in the oceans due to their warmer surface temperatures, increased water vapor (also a greenhouse gas) in the atmosphere due to warmer air, and there are others. Negative feedbacks are also occurring, such as the increased reflectivity of clouds formed by the higher atmospheric water vapor content, increased global photosynthesis, and increased shading by atmospheric pollution and dust from deforestation and expanding deserts. It is the relative magnitude and interactions of these feedback loops as well as the timing of their maximum impacts that make long-term predictions of global climates so difficult. However, the consensus is that the positive feedbacks will far outweigh the negative feedbacks in influencing future global climate.

Evidence from the Past

So atmospheric CO_2 levels are increasing and will most likely continue to increase in the future. What will be their effect? One approach to predicting the future is to study the past. What was the Earth like in the past when there were naturally elevated atmospheric CO_2 levels? By measuring CO_2 levels in air bubbles trapped in the ice of dated glacial cores, we know that around 35 million years ago the atmosphere contained about 1,000 ppm CO_2. This is close to where we'll be in 2100, assuming that there have not been Herculean efforts in the interim to reduce CO_2 release into the atmosphere. By studying evidence from the biological communities in ocean sediments dated around 35 million years ago, we know that the average ocean temperatures at that time were about 5 to 10 °C (9 to 18 °F) warmer than they are today, and at the polar latitudes they were 25 to 30 °C (77 to 86 °F) higher than current temperatures.[5] These are ocean temperatures—summer terrestrial temperatures at these same latitudes would be much higher.

Although most climate models predict somewhat lower global temperature increases by the end of this century, the reality is that we don't really know the magnitude of the increase. We are in the midst of a global experiment with an unknown outcome, and there is no experimental control planet to retreat to if this experiment goes awry. The problem is real, and the sooner our generation takes it seriously, the more habitable the planet will be for our grandchildren and their families.

Observed Effects
Meanwhile, back in the present, what will be the effect of global warming on the Adirondacks and our lakes and watersheds? One thing we know is that the ranges of some plants and animals are shifting north. For the last 40 years, the Audubon Society has conducted a Christmas Bird Count during which thousands of volunteers throughout North America observe birds for one 24 hour period within a few days of Christmas. This is the most long-term, wide-ranging and reliable data set describing bird populations in North America. Analysis of these data indicates that 58% of the 305 widespread bird species that overwinter on the continent have significantly shifted their ranges northward since 1968. For instance, the range of the red breasted mergansers has shifted north by about 250 miles, and the American black duck range shifted by 140 miles.[6] The Cornell Ornithology Laboratory's Project Feeder Watch program has also documented the northward shift of bird ranges, with the red-bellied woodpecker becoming much more common in New England in the last two decades.[7]

Global warming may be considered a good thing if you live in Saskatchewan, love red-bellied woodpeckers or hate long winters, but the ranges of more troublesome species are also shifting northward. For instance *Hydrilla*, an invasive aquatic weed from the warmer regions of Asia, is potentially more of a problem to our lakes than milfoil because it grows more densely. It was first established in Florida in 1960 and has typically been considered a warm-water species that would not be a threat to the northeastern U.S. However, it was discovered for the first time in New York in 2008 and is apparently slowly marching northward.

A terrestrial counterpart, the hemlock woolly adelgid (*Adelges tsugae*), is an aphid-like insect that sucks the juices out of eastern hemlock trees and eventually kills them. It was introduced into Virginia from southern Japan in the 1950's, and it quickly spread throughout the southern reaches of the hemlock range. It is now moving northward, and since it is a terrestrial creature that must overwinter, the assumption is that it will eventually be limited in its northward march by frigid winters. However, winters are not as frigid as they once were, and it reached Albany County, New York, in 2003 and expanded into southern Vermont in 2008. I experienced first hand the effects of this nasty beast when I fished in central Pennsylvania in the summer of 2007. In all the beautiful, isolated stream valleys we fished, virtually all the hemlocks were infested, and many large trees had already died. I can't imagine what these valleys would look like in another decade or so if all the hemlocks die. Is that what we have to look forward to in the Adirondacks?

With recently introduced invasive species such as *Hydrilla* and the wooly adelgid, it is hard to know for certain whether a northward range shift is due to the climate warming or to the natural range expansion of a species that has yet to reach its ecological limits. However, all species have an ecological niche limited at some level by cold temperatures, and the geographical range of all those species will shift to the north as the climate warms.

Future Effects

Native plant ranges will also move northward. If the long range predictions of temperature increase hold true, the range of sugar maple will eventually be pushed out of the northeastern U.S. and into Canada. So much for New England's famed maple sugar industry and the annual autumnal leaf-peeper tourist migrations.

Another result of warmer, wetter weather in the Adirondacks will be longer dry periods, though more severe storms when they do occur.[8] In highly developed or agricultural watersheds, this means more flooding, since the natural vegetation and wetlands are no longer present to act as buffers to slow the runoff as it flows into the streams. In wooded watersheds such as those in most of the Adirondacks, the increase in stream flow from a storm will not be as rapid, nor will it carry the sediment load of the developed watersheds. But the enhanced dry–wet cycles will no doubt result in greater fluctuations of our lake levels with their associated problems of water clarity, shoreline flooding and access, and dock maintenance.

A more direct effect of future warming on our lakes will be higher summer water temperatures and longer ice-free periods. These will result in inevitable shifts in species composition—some predictable, though many not—and a longer growing season for aquatic plants. In addition, the longer time period between turnovers will allow the hypolimnion more time to become, or to be, anoxic. In the case of Sacandaga Lake, instead of a brief period of low dissolved oxygen levels in the hypolimnion in late September, it could last for many weeks with resultant negative effects on the coldwater fish populations like trout. A similar effect will be found in streams, since warmer stream temperatures will also affect trout fishing and survival.

Other Concerns

We can also be sure that the future will hold a continuing battle with invasive species, as discussed in Chapter 11. Some will be impossible to stop, like the spiny water flea, while others, like zebra mussels, may not be able to survive in many Adirondack lakes that have levels of calcium too low for their growth. The spread of vascular invasive plants such as milfoil can be slowed by inspections of boats and trailers at boat ramps, but this requires staffing at the ramps and education of the boating public.

Another future threat to Adirondack lakes is nutrient loading due to shoreline development. The removal of shoreline forests increases sediment transport and nutrient enrichment of adjacent lakes, and the paving of previously unpaved areas increases high-nutrient runoff into nearby lakes. These problems can be mitigated by practicing responsible lakeshore stewardship, such as those described in previous chapters and in Appendix V.[9] Despite more stringent state regulations on septic tank and tile field locations, the expansion of sewer lines to shoreline properties, wherever possible, should be encouraged.

Added to these issues is continuing habitat destruction from human development projects, budget-driven reductions in state regulatory agents and programs, and political pressure to reduce the role of government in rural areas. Also we have only recently become aware of the effects of some low-level water contaminants not previously understood. Discarded and excreted pharmaceuticals, as well as chemical endocrine disruptors like PCB's (polychlorinated biphenyl) have been found to change sex ratios in natural aquatic animal populations downstream from wastewater treatment plants.[10] These agents can be active at extremely low concentrations—at the parts per billion and even parts per trillion levels. What will be the long-tern effects of these chemicals on communities that use surface water for consumption?

The list goes on.

Any Good News?

However, all the news is not bad. Air quality has improved due to the Clean Air Act and its amendments, and the water quality of many Adirondack lakes has improved. But we must remain vigilant to maintain and even strengthen the restrictions on atmospheric mercury, NOx, SOx and other toxicants. Considerable scientific research has better defined many environmental problems so that solutions can be developed. In addition, the population in general is more familiar with environmental issues, and in most cases is supportive of environmental legislation. We must all do our part to encourage these trends.

As Niels Bohr pointed out in the introduction of this chapter, we can never really be sure what the future holds. Who, for instance, could have predicted 50 years ago the widespread effects of acid rain, or 20 years ago the presence and negative effects of environmental mercury? All we can do is to learn as much as we can about our "Earth's eyes," as Thoreau once described our lakes, and carefully monitor them so that we can quickly notice if something is amiss. We must encourage an environmental ethic that discourages simple, thoughtless actions such as emptying a bait bucket into a lake, not cleaning a boat or trailer of vegetative matter before putting it into a lake, or purchasing animals or plants well known to be invasive. The time of passively enjoying our lakes is past. We must all contribute to the effort of their maintenance, and even restoration, if we are not to have a destiny for our lakes as implied by the NY Yankee player/manager Yogi Berra: "The future ain't what it used to be."

Appendices

Appendix I
MEASUREMENT UNIT CONVERSIONS

Virtually all scientific research uses the metric system of measurement. Within the English systems of measurement (whether U.S. or U.K./Imperial), it can be awkward to convert from length to volume. For example, how many cubic inches of water are there in a quart? Similarly, conversions of length to area (43,560 square feet per acre?) or even one length to another (5,280 feet per mile?) within English systems can be a challenge. However, the metric system is designed to make these conversions easy. Here are some common units of conversion:

Metric System Conversions
1 meter (m) = 100 centimeters (cm) = 1,000 millimeters (mm)
1 cm = 10 mm
1 mm = 1,000 microns (μm)
1 cm^3 water = 1 gram (g) = 1 milliliter (mL)
1 g = 1,000 milligrams (mg)
1 liter (L) = 1,000 mL = 1,000 g = 1 kilogram (kg)
1 part per million (ppm) = 1 mg per liter (mg/L)

Metric to English System Conversions
1 cm = 0.4 inch (2.5 cm = 1 inch)

1 m = 3.3 feet = 1.1 yard = 39.4 inches
1 g = 0.034 ounce
1 liter = 1.06 quart

Temperature

Temperature is typically measured using either the Centigrade (°C) or Fahrenheit (°F) scale. The freezing point of water is 0 °C (32 °F), and its boiling point is 100 °C (212 °F). Here is a quick conversion table:

Table I.1
Conversion of Centigrade to Fahrenheit temperature systems.

Centigrade (°C)	Fahrenheit (°F)	Centigrade (°C)	Fahrenheit (°F)
0	32	25	77
5	41	30	86
10	50	35	95
15	59	40	104
20	68		

Quick Estimates

Here are a few rules of thumb worth committing to memory:

- ✓ A meter is just a little longer than a yard.
- ✓ A liter is just about a quart.
- ✓ A millimeter is the same thickness as a dime.
- ✓ A gram is a little more than the weight of 2 standard paper clips.
- ✓ A nickel weighs five grams.
- ✓ A nickel is two cm in diameter.
- ✓ The diameter of a human hair ranges from about 40 μm (blond) to about 150 μm (black), with an average of about 100 μm (or about 1/10th of a millimeter).

Appendix II
More About pH

THE ACIDITY OF A SOLUTION is determined by the concentration of its hydrogen ions (H^+). Solutions with high H^+ concentrations are said to be "acidic," whereas those with fewer H^+ are more "basic." The number used to describe a solution's H^+ concentration is called its pH, and it is calculated as the negative logarithm of its H^+ concentration. Although this definition may seem intimidating to the less scientifically inclined, it simply means that for each difference in pH number value (for example, from pH 6 to pH 7), the concentration of hydrogen ions changes by a factor of ten (hence the logarithm). Also, solutions with lower pH values have higher hydrogen ion concentrations than solutions with higher values (hence the negative logarithm). So a solution with a pH of 6 has ten times *more* H^+ than solutions with a pH of 7.

Pure water (H_2O) will naturally dissociate into H^+ and OH^- at a level of pH 7, which is considered a "neutral" pH. Solutions with a lower pH than 7 are considered more acidic, whereas those with a pH greater than 7 are called more basic. Compounds added to pure water can either add hydrogen ions to the solution (reducing its pH) or subtract them (increasing its pH). Figure II.1 shows the pH values for some familiar liquids.

When carbon dioxide (CO_2) is dissolved in water, a chemical reaction occurs that produces carbonic acid, thereby increasing the H⁺ in the water and decreasing its pH to about 5.6. As discussed in Chapter 11, acid rain also increases the concentration of H⁺ within lakes.

Other compounds will decrease the H⁺ concentrations, and still others will tend to keep the pH of a solution relatively constant. These latter compounds are called "buffers," and a sophisticated system of buffers keeps the pH of human blood at about 6.8. Buffers are also important to natural systems. One such natural buffering system involves calcium carbonate ($CaCO_3$), the major component of limestone and its metamorphosed

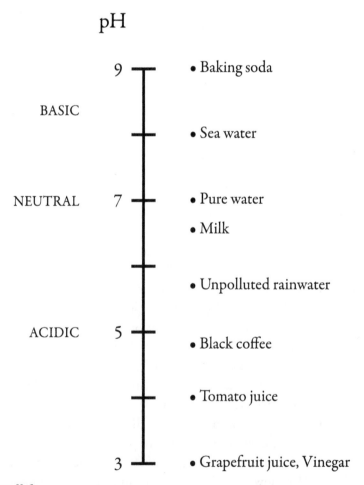

Figure II.1
The pH scale, including values for some familiar liquids.

cousin, marble. The presence of limestone or marble in a lake's watershed can compensate for (neutralize) acidic precipitation and runoff from naturally occurring sources of acidity, allowing a lake to keep its pH within the neutral range where biological activity is the most successful. More about the biological impact of acid waters on a lake's ecosystem can be found in Chapter 11.

Appendix III
More About Weird Water

As discussed in Chapter 3, we would expect that liquids containing molecules with a greater mass would boil at higher temperatures than those with less mass, since more energy would be required to speed up these heavier molecules fast enough to allow them to break free from their liquid neighbors to form a gas. Table III.1 summarizes boiling point data from the beginning of Chapter 3.

Note that water requires well over 400 °F more heat (energy) to vaporize than similar sized molecules (for example, −260 °F for methane versus

Table III.1
The boiling point for selected molecules, relative to their size.

Molecule	Chemical Formula	Molecular Mass	Boiling Point (°F)
Methane	CH_4	16	−260
Water	H_2O	18	+212
Ethane	C_2H_6	30	−126
Hydrogen Sulfide	H_2S	34	−80

Note. Molecular mass is measured in grams per mole (g/mol).

+212 °F for water). To explain this boiling point anomaly, it is important to understand a little about bonds both within the water molecule and between adjacent molecules.

Water consists of two hydrogen atoms attached to an oxygen atom by bonds called *covalent bonds* (Figure III.1). These are common bonds that link the atoms in most molecules, including all the molecules listed in Table III.1. So what makes the bonds linking atoms in water different? In all covalent bonds, electrons are shared between the bound

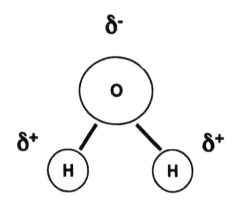

Figure III.1
A water molecule (H_2O). Note the slight charge on the atoms due to the strong electron affinity of the oxygen atom relative to the hydrogen atoms. The oxygen is negative relative to the two hydrogens.

atoms, making the bond very strong. In most cases, the electrons are shared about equally between the two atoms. However, oxygen has a much stronger affinity for electrons than most other atoms, especially hydrogen. So in water, the electrons are not shared equally but tend to stay more in the vicinity of the oxygen atom. Since electrons have a negative charge, the oxygen atom is relatively more negatively charged than the hydrogen atoms. This allows an *inter*molecular attraction between the negatively charged oxygen atoms of one water molecule, and a positively charged hydrogen atom of an adjacent water molecule. These bonds are much weaker than covalent bonds, and are called *hydrogen bonds* (Figure III.2). None of the other molecules in Table III.1 have hydrogen bonding, since carbon and sulfur have electron affinities that are much weaker than oxygen. It is these intermolecular hydrogen bonds between the water molecules that make its boiling point so much higher than other molecules of similar size.

To explain the water density anomaly discussed in Chapter 3, in which water is at its most dense at 4 °C (See Table 3.1), we must look further into the interactions of the water molecules and their hydrogen bonds. Molecules in a liquid form are in constant motion. When heated, they move more quickly (so that the liquid expands and its density decreases), and

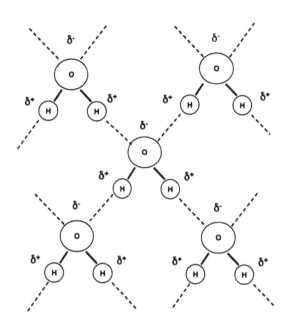

Figure III.2
The lattice structure in water (H_2O) created by hydrogen bonds (dashed lines).

when cooled, they move more slowly (and the liquid's density increases). However, when water cools down, the velocity of its molecules slows and hydrogen bonding becomes more prevalent. As more and more of the water molecules form hydrogen bonds with their neighbors, they form a kind of lattice—the colder the water gets, the higher the proportion of its molecules is involved in this lattice (Figure III.2). At 0 °C (32 °F), virtually all the molecules are in the lattice and water forms a solid. But the molecules in the lattice are not as dense as they were before the lattice formed, so there are some spaces within the lattice where non-bound water molecules can fit. Therefore, at 4 °C (39 °F) there are many molecules in the lattice, but also many within the interstices of the lattice, so that water is at its most dense at this temperature. As these interstitial water molecules join the lattice when the water cools further, the density of water decreases. When its density decreases, its volume expands, and as ice forms, it will break glass containers filled with freezing water and crush docks and other structures left in a lake in the winter. This phenomenon does not occur in any other liquid.[1] The result of this temperature–density curve is a critical factor in understanding how a lake functions. Because water is at its most dense at 4 °C (39 °F) and ice floats, a winter lake will always have liquid water just below the ice, extending all the way to the lake bottom.

The lattice illustration in Figure III.2 is simplified so that it will fit on a two dimensional page. In reality, two hydrogens from two different water molecules will be bound with each oxygen, and the angles of the bonds produce a hexagon, giving snow flakes and individual crystals in the ice sheet of a winter lake their hexagonal shape.

Appendix IV
Taxonomy & Scientific Classification

THE SCIENCE OF NAMING ORGANISMS has been known as *taxonomy*, but it is much more than that. The goal is to develop a system that relates organisms in a way that reflects their evolution. This science is now known as *systematics*. The classification hierarchy going from the largest, most inclusive groups to the smaller, most closely related groups consists of the following categories: Kingdom, Phylum, Class, Order, Family, Genus, and Species.

In the above hierarchy, a category above is always more inclusive than the one below. There are many Phyla in a given Kingdom, many Classes in a Phylum, and so on. Carl Linnaeus, the 18th century Swedish scientist, contributed the concept of binomial nomenclature, used to name a given species. It consists of a *genus* name, always capitalized, and a *species* name, rarely if ever capitalized. Since both names are usually derived from Greek or Latin, they are italicized in a text.

If we were to use the Linnaeus binomial system to classify us, it would be the familiar *Homo sapiens*. The more expansive systematic hierarchy would look like this:

Kingdom: Animalia (heterotrophic, ingests food)
　Phylum: Chordata (having a dorsal nerve cord)
　　Class: Mammalia (warm blooded, mammary glands, body hair)
　　　Order: Primates: (opposable thumbs)
　　　　Family: Hominidae (large braincase, similar dentition)
　　　　　Genus: *Homo* (walking upright, bipedal)
　　　　　　Species: *sapiens* (the "wise" homo)

Since most of the species in a given genus are similar (*Homo habilis* and *Homo erectus* both look pretty similar to *Homo sapiens*), for simplicity's sake in this book, I typically use just the genus designation to name a creature. By the way, the plural of "genus" is "genera," whereas the plural of "species" is identical to its singular, also "species."

Appendix V
A Primer on Lake Stewardship

THERE ARE SOME VERY GOOD published guidelines available for local lake stewardship, and some are listed at the end of this appendix. An excellent one is *Diet for a Small Lake*[1], published jointly by New York State Department of Environmental Conservation (NYSDEC)[2] and the New York State Federation of Lake Associations (NYSFOLA)[3]. Information and strategies in *Diet* are relevant to all temperate lakes, not just those in New York State. Other states may have comparable publications. *Diet* is far more comprehensive than anything I can produce here, so I will just provide an overview of some of the things individuals or organizations can do start monitoring a lake.

Water Quality

It is very difficult to detect degrading water quality if you don't have some background information on your lake for comparison. In New York, there are a few organizations that regularly collect data on lakes and have been doing so for years. The Citizen's Statewide Lake Assessment Program (CSLAP), run by NYSDEC, trains volunteers throughout the state to collect water samples that are then analyzed by certified labs. The Adirondack Lake Assessment Program (ALAP) uses volunteers and professionals to collect data on Adirondack lakes. In Hamilton County, the Soil and Water Conservation Service (SWCS) collects lake data on county lakes. These programs are mainly complementary, though there is some overlap. So before you start a program on a given lake, be sure to search the archives of these organizations to see if your lake is among those sampled. Other states may have similar programs with accessible data.

Any one of these or similar programs is vulnerable to the whims of financial uncertainty or politics. Whether your lake has been included or not, it is best to start sampling your lake soon. The more detailed and long-term your baseline data, the sooner you can detect problems. So start establishing a monitoring scheme even if you are currently satisfied with your lake's water quality.

Water Transparency

Water transparency is a good indicator of the nutrient status of a lake. In most northern lakes, the size of algal populations is most frequently the cause of reduced water transparency, and algae grow in response to enriched nutrient levels. The most common method of measuring water transparency is the Secchi disc. A simple homemade Secchi disc is cheap and, if used regularly, can yield valuable data on water quality. You can also purchase one for about $30 to $50 at biological supply houses (Carolina Biological, Cole Palmer, Ben Meadows, among others).

Secchi depths should be recorded as described in Chapter 3 at regular intervals throughout the season, usually at the deepest site of the lake. Sampling along various shorelines and bays may indicate "hot spots" of nutrient input. The longer the duration of the data set, the more reliable it will be, since yearly variations will tend to mask real trends. For example, the Secchi depths in Lake Tahoe have shown a marked decrease of almost 40% since 1964 (from 95 ft to 60 ft), indicating a definite problem with cultural eutrophication. However, selected five-year periods within that longer 56-year period suggest that water transparency was actually improving. So a short data set of only a few years can be very misleading.

Water Chemistry

You can also measure levels of specific chemicals in your lake, but this requires a more significant expenditure. A water testing kits can cost from $200 to $600, depending on which chemicals you want to measure. Hach

Company and LaMotte are two such suppliers, but there are many others. The following are some of the chemicals you may be interested in measuring.

Calcium levels can suggest whether your lake is vulnerable to zebra mussel invasion. The threshold seems to be about 15 to 20 ppm. Calcium levels also indicate the amount of limestone or marble in your watershed, which in turn indicate your lake's resistance to acidification. A long-term decrease in calcium levels may indicate a reduction of your watershed's acid neutralizing capacity (ANC), possibly a warning of future acidification.

Chloride levels nowadays mainly indicate the presence of road salt runoff, but they can also point to sources of wastewater contamination or septic seepage. It's hard to separate the two, unless you find a "hot spot" of localized elevated levels.

Lake *pH* measures acidity, and levels are usually in the range of 6 to 8. It is related to both the concentration of acidic precipitation and the watershed's ANC. A pH lower than 6 may explain why you have been having so much trouble catching fish in your lake.

Nitrate levels can vary dramatically over the growing season (see Figure 4.4), so their interpretation is difficult, especially since all the inexpensive tests measure *dissolved* nitrate, and much of the nitrogen in a lake is concentrated in its organisms.

Chlorophyll a levels indicate the size of algal populations, and some inexpensive tests can measure this. But reliable *phosphorus* (P) levels, the major limiting factor in most lakes, are more difficult to measure. The relevant measurement is *total phosphorus* (TP), which includes both dissolved P and P incorporated into organisms. The relevant levels of both chlorophyll *a* and TP are in the parts per billion (ppb or µg/L) range, and inexpensive tests are not capable of giving reliable readings at these low concentrations. Certified commercial labs are the best source of reliable data.

Inexpensive *conductivity* meters can cost less that $100 and will measure all the ions in the water. Although they cannot distinguish specific chemicals, they can indicate wastewater contamination as well as sites of runoff outfalls in a lake. In areas where hydrofracking is practiced, an increase in conductivity may indicate contamination with the brine used in the process.

Aquatic Plants

Learn which aquatic plants are in your lake. The three published sources listed at the end of this appendix are all excellent for learning about both native and invasive species. *Through a Looking Glass: A Field Guide to Aquatic Plants*[4] is a good place to start. You will find excellent drawings and photos, developed specifically for volunteer lake monitors, in the *Maine Field Guide to Invasive Aquatic Plants, and Their Common Native Look Alikes*[5]. The most comprehensive treatise on aquatic plant identification can be found in *Aquatic and Wetland Plants of Northeastern North America*[6]. This is a two volume set and not recommended for the beginner.

If you make a rough map of the species distributions in your lake, you would be able to see whether they are expanding or contracting in the future. To zero in on invasive species, explore the Web site of the Adirondack Park Invasive Plant Program (APIPP)[7].

If you do find invasive species, hand-pulling, sediment mats, or harvesting by cutting stems or raking are all methods that can be done by individuals in small areas, for instance around docks. However, check with statewide regulations because you may need a permit to clear larger areas. If you use improvised sediment mats (such as tarps or rugs), make sure that they are well weighted and vented, because sediment degassing will produce gas bubbles that will otherwise make the mats float.

Large areas or lake-wide treatments will require state permits and the use of certified contractors, in most cases. Various larger scale control techniques are listed in Chapter 11, along with their costs and possible outcomes.

Lakeshore Management

Paved portions of a watershed as well as lawns are major sources of nutrients that enter a lake (see Chapter 11). Extensive lakeshore lawns should be discouraged while shorelines strips of natural vegetation encouraged. Storm water runoff should either be diverted from our lakes, or captured in settling basins established in the runoff flow prior to lake outfall. A good guide to managing lakeside land is *Landscaping for Wildlife and Water Quality*[8].

This brief summary is not meant to be comprehensive—just enough to give you a rough idea of some of the things you might do to initiate a monitoring program at a lake and some sources for a more thorough coverage of these topics. Since the most valuable aspect of the data you collect is to show trends, it is important that these data be kept in a form that can be passed on to future lake monitors. Lake associations should provide this type of continuity, or perhaps local planning groups or your local soil and water conservation service. Good luck, and take pride in the knowledge that you are contributing to the preservation of one of our greatest natural resources.

A Few Sources to Assist in Lake Stewardship

Here is a good starter list of sources to help you understand, evaluate, monitor, and maintain the quality of your lake. There are many others. Check with your state environmental offices.

General Stewardship

1. NYSFOLA and NYSDEC. 2009. Diet for a small lake: the expanded guide to New York State lake and watershed management. 2nd ed. 352 pp.

2. New York State Department of Environmental Conservation (NYSDEC). Address: 625 Broadway, Albany, NY 12233-0001. Web site: www.dec.ny.gov.

3. New York State Federation of Lake Associations, Inc. (NYSFOLA). Address: Post Office Box 84, LaFayette, NY 13084-0084. Web site: www.nysfola.org.

Aquatic Plants

4. Borman, S, R Korth, and J Tente. 2001. Through the looking glass: A guide to aquatic plants. Wisconsin Dept. Nat. Res. Publ. No. FH-207-97.

5. Maine Center for Invasive Aquatic Plants (MCIAP). Maine field guide to invasive aquatic plants, and their common native look alikes. 2007. Address: Maine Volunteer Lake Monitoring Program, 24 Maple Hill Road, Auburn, ME 04210-8774. Web site: www.mainevlmp.org.

6. Crow, EG, and CB Hellquist. 2000. Aquatic and wetland plants of northeastern North America. 2 volumes. University of Wisconsin Press.

7. Adirondack Park Invasive Plant Program (APIPP). Address: Post Office Box 65, Keene Valley, NY 12943-0065. Web site: www.adkinvasives.org.

Lakeshore Management

8. Henderson, CL, CJ Dindorf, and FJ Rozumalski. 2009. Landscaping for wildlife and water quality. Minnesota Dept. of Natural Resources. 176 pp. Address: 500 Lafayette Street, St. Paul, MN 55155-4040. Web site: www.dnr.state.mn.us.

Glossary

A

Acid neutralizing capacity (ANC) – The ability of a soil to reduce the acidity of water filtering through the soil. ANC is usually associated with limestone or marble in the watershed.

Akinete – An enlarged cell in filamentous blue-green algae used to store nutrients and act as a survival structure that enables the alga to overwinter and survive periods of desiccation.

Algae – A diverse group of small (usually microscopic) photosynthetic aquatic organisms that provide the basis for most aquatic food webs.

Anoxic – Lacking oxygen.

Anthropogenic – Created by humans.

Aphotic zone – The deeper zone of lake where there is so little light penetration that photosynthesis is impossible.

Autotrophs – Organisms that can produce their own food, usually by photosynthesis.

B

Benthos – The bottom, solid layer of an aquatic system. In a lake, it's the sediment.

Bioaccumulation (biomagnification) – The process of concentrating a compound in organisms in the higher trophic levels in a food web. The compound must be metabolically stable and not be excreted by the organisms. It is usually sequestered in tissues such as fat or bones with low turnover rates.

Biomass – The total mass (similar to weight) of all the living biological organisms in an area.

Black ice – The first ice formed on a body of water in early winter that is very strong and transparent. It looks "black" because the water below it appears dark.

Blue-green algae – Prokaryotic cells that use chlorophyll *a* as a photosynthetic pigment to produce sugars and oxygen. They appear blueish due to pigments other than chlorophyll. They are also call cyanobacteria.

Boreal forest – The most northern forest type, just south of the tundra, that consists mostly of conifer trees and a few hardwoods such as aspen.

Bottom-up control – The control of a given trophic level in a food web by manipulating the levels below it. For example, controlling algae by limiting nutrients.

Buffer (chemistry) – A chemical compound that can take hydrogen atoms out of a solution to reduce the solution's acidity. Due to a series of reactions, a buffer tends to stabilize the solution's acidity at a consistent pH. In general: Something intermediate between two opposing entities that modulates their effect on each other. For example, a native vegetation buffer between a lawn and a lake that decreases the negative effect of the lawn on the lake.

C

Caddisflies – The group of insects having an extended aquatic immature stage that often builds a protective structure of sand, sticks or other vegetation. Its adult stage has long antennae and wings that fold like a tent when at rest.

Carapace – The clear, shell-like covering of the body of water fleas (Cladocerans).

Carnivores – Animals that eat other animals.

Character displacement – A mechanism for reducing competition for organisms competing for the same limited resource. Physical differences in competing populations are amplified through natural selection so that competition is reduced.

Ciliates – Unicellular animals covered with many small, whip-like structures used for locomotion and feeding.

Cladocerans – Small (± 1 mm) filter-feeding animals with two large anterior swimming appendages, a clear body covering and a dorsal brood chamber for developing embryos. Also called water fleas.

Clear lake phase – A period (usually in the late spring) in eutrophic lakes when water transparency temporarily improves, caused by increased predation of the algae by filter-feeding zooplankton.

Clovis point – A stone spear point made with a characteristic flaking pattern. It is named for the Paleo-Indian culture of Central and North America that dates from 13,000 years ago, but more recently in the northeast.

Compensation point (level) – The depth in a lake where the light intensity is decreased to the level where the biological community photosynthesis equals respiration—the level where there is no net production of oxygen.

Competitive exclusion principle – The concept that two species competing for the same limiting resource cannot continue to coexist. Either one will be driven to extinction, or natural selection will alter one or both to reduce competition.

Consumers – Organisms that cannot make their own food. They rely on gaining nutrients by ingesting other organisms or organic material produced by autotrophs. Also called heterotrophs.

Continental drift – The movement of continental land masses perched on top of the rigid plates as described by the theory of plate tectonics.

Copepods – Small aquatic animals, usually 1 to 2 mm long, with two large anterior swimming appendages and a single eye. They may be filter feeders or predators on other animals.

Covalent bonds – The bonds between atoms in which electrons are shared. They are very strong.

Cultural eutrophication – The addition of nutrients to an aquatic system by human activity.

Cyanobacteria – Prokaryotic photosynthetic organisms, often called blue-green algae. They produce oxygen as a product of photosynthesis.

D

Decomposers – A type of consumer that lives on dead organic matter. They recycle inorganic nutrients in an ecosystem.

Diatoms – A type of eukaryotic algae that has a silicaceous cell wall, often elaborately ornamented.

Diffuse source (of pollution) – Pollution originating from a large area with no localized source.

Diploid – Containing a full complement of chromosomes—one set from each parent.

E

Emergent plants – Vascular plants that have much of their tissue above the surface of the water. Found in the shallower parts of a lake.

Ephippium – The modified portion of the cladoceran carapace containing the "resting eggs" (in actuality, fertilized zygotes). It protects the "eggs" and facilitates dispersal.

Epilimnion – The top layer of a stratified lake above the thermocline. It is the warmer part of the lake, is mixed by the wind, and readily exchanges oxygen with the atmosphere.

Eukaryotes – Organisms with a true membrane-bound nucleus and organelles such as mitochondria, chloroplasts and others.

Euphotic zone – The upper portion of a lake where light intensity is high enough so that photosynthesis is possible. Often defined as the volume of the lake above the compensation level.

Eutrophication – The addition of nutrients to an aquatic system. It can be a natural process, as with leaves and other material being washed into a lake, or it can be the result of human activities, called cultural eutrophication.

F

Fault (geologic) – A break in the earth's crust due to movement of the crust.

Feedback – A situation in which the result (output) of a process influences the rate of production of that output. For example, in a positive feedback loop, the formation of ice at the Earth's poles increases the reflectivity of the Earth, cooling it and thereby increasing the rate of formation of ice. In a negative feedback loop, the output decreases the rate of formation of that output.

Flagellates – Types of Protozoa that contain one or a few long whip-like structures used for locomotion.

G

Green algae – A type of eukaryotic algae that contains chlorophylls *a* and *b*, has cellulose cell walls and stores food as starch.

H

Haploid – Having only one set of chromosomes. In animals the sex cells (sperm and egg) are the only haploid cells. Plants have more complex life cycles.

Herbivores – Animals that eat autotrophs.

Heterocyst – The enlarged, thick-walled cells in some filamentous blue-green algae that are associated with nitrogen fixation.

Heterotroph – Organisms that cannot produce their own food. They must ingest or absorb complex organic molecules (their food) from their environment.

Hydrogen bonds – Weak intermolecular bonds between a hydrogen atom on one molecule and an electronegative atom (such as oxygen) on an adjacent molecule.

Hypolimnion – The cooler bottom layer of a stratified lake below the thermocline. It is isolated from the atmosphere and the epilimnion by the thermocline.

I

Ice candles – Long, thin hexagonal crystals of ice that are the result of ice melt on a lake. The water between the crystals melts first as the ice mass warms, freeing the individual crystals.

Internal loading of a nutrient – The recycling of nutrients into the water column from the sediment within a lake.

Isothermal – Having the same temperature. In a lake, this occurs during the spring and fall when the entire water column is the same temperature.

K

Kettle lakes – Lakes that are formed as a glacier retreats. The retreating glacier leaves large blocks of ice that are then surrounded by glacial outwash debris. When the ice block melts, a depression remains, resulting in a kettle lake when it fills with water.

L

Lacunae – A cavity in the stem of aquatic plants that assists in the transport of gasses among the leaves, stems and roots.

Larva – An immature stage in the life cycle of an insect that requires a resting pupal stage before becoming an adult. It often looks like a caterpillar.

Lateral line (in fish) – A sense organ along the approximate median line on the side of a fish that senses movement and vibrations in the surrounding water.

Limestone – A sedimentary rock that contains mostly calcium carbonate. It is important in a watershed because it can buffer acids (reduce acidity) in rainwater as it percolates through the soil on its way to a lake.

Limiting nutrient – The nutrient that will limit the growth of an organism or population of organisms.

Limnology – The study of the physical, chemical and biological aspects of inland water bodies, typically fresh water.

Littoral zone – The shallower portion of a lake where rooted vascular plants can survive. It is related to water transparency—the more transparent the water, the deeper the littoral zone.

Lorica – The hard transparent shell of a rotifer.

M

Macro algae – A group of plants with very simple structure—one genus has a single elongated cell as its "stem." Because of this simple structure, they were originally considered complex algae, but are now put into a taxonomic group of their own. The name "macro alga" is a relic of this earlier classification, but is still commonly used.

Macro invertebrates – The group of insects that have an extensive immature aquatic stage and relatively short adult stage. It includes the mayflies, stoneflies and caddisflies.

Marble – A form of metamorphosed limestone. Like limestone, it consists mainly of calcium carbonate and reduces the acidity of water in the soil.

Mayflies – A group of insects having an extended aquatic immature stage in their life cycle and a brief adult stage. Aquatic nymphs typically have three tail-like structures and abdominal gills. Their short-lived adult stages fold their wings vertically, like a sail above their body when at rest.

Meiosis – A type of cell division that reduces the number of chromosomes in the resultant cells by one half. In animals, it produces the egg or sperm.

Mesotrophic – Describes an intermediate nutrient status of a lake, between a nutrient-rich eutrophic lake and a nutrient-poor oligotrophic lake.

Metalimnion – The region in a lake between the epilimnion and hypolimnion that includes the thermocline.

Methyl mercury – A form of mercury produced from elemental mercury in the anoxic sediments of aquatic systems by bacterial action. It can bioaccumulate and is considered toxic at relatively low levels. Effects include neuropathy and reduced reproduction in affected animals.

Micron (μm) – One millionth of a meter (10^{-6} m) The same as one thousandth (10^{-3} mm) of a millimeter.

Mid-Atlantic ridge – The raised portion of the seabed about midway between the Americas and Europe and Africa. It has extensive seismic activity, with the resultant magma deposits spreading the seafloor and pushing the Americas away from Europe and Africa.

Midge – A type of insect with an extensive aquatic immature stage and an adult stage that looks like (but is not) a mosquito.

Mitosis – The type of cell division that produces exact genetic replicas as daughter cells.

Moraine (glacial) – A glacial debris deposit at either the sides (lateral moraines) or front (terminal moraines) of an advancing glacier.

N

Nauplii – The immature stage of a copepod.

Niche – The description of an organism's position and function in an ecosystem, including its nutritional and habitat requirements.

Northern hardwood forest – The forest type typical of much of New England, with sugar maple, beech, hemlock and yellow birch as prominent constituents.

Nymphs – The immature stage of an insect's life cycle that transforms directly into an adult.

O

Oligotrophic – Describes the nutrient status of a lake with low nutrient levels. Water transparency is high, phosphorus levels low, and algal populations low.

Orogeny – The formation of mountain ranges as the result of the interactions of continental plates.

P

Pangea – The name given by Alfred Wegener to the supercontinent that contained all the continental plates before it split apart about 200 million years ago.

Parthenogenesis – A method of reproduction during which the female produces diploid "eggs" by means of mitosis. There is no sexual reproduction. The "eggs" are really embryos that are genetically identical to the mother. It is a method of cloning used by rotifers and cladocerans that allows a rapid expansion of the population.

Parts per million (ppm) – One unit of a substance found within one million units of another substance, typically water.

Pelagic zone – The deeper portion of a lake found outside of the littoral zone. In most lakes, it contains most of the lake's volume.

Petiole – The stem-like projection at the base of a leaf blade that connects the leaf to its stem.

pH – A unit describing the acidity of a solution. It measures the hydrogen ion concentration on a logarithmic scale.

Phytoplankton – Plankton that can carry out photosynthesis.

Piscivore – An animal that eats fish.

Planktivore – A fish or other animal that eats plankton.

Plankton – A diverse group of aquatic organisms that are weak swimmers.

It includes algae, protozoa, rotifers, cladocerans, copepods and many aquatic insects.

Plate tectonics – The theory that the outer layer of the Earth is divided into large rigid plates that move and interact at their edges. They float on the viscous mantle below, and move by means of convection currents within the mantle.

Pleistocene – A period in geological history starting about 2 million years ago characterized by cyclic global glacial activity.

Point source (of pollution) – A source of pollution that can be localized. Point sources are much more easily controlled than diffuse sources.

Producers – Organisms (usually plants and algae) that can produce complex organic compounds (such as sugars) from carbon dioxide and an external energy source such as light.

Production (ecological) – The total amount of photosynthetic product formed per area of an ecological system. Can be measured in units of biomass or energy.

Prokaryotes – Organisms that contain no membrane-bound organelles, such as a nucleus, mitochondria or chloroplasts. Bacteria and blue-green algae are prokaryotic.

Protozoa – Single-celled animals.

Pseudopodia – An extension of the cytoplasm of an amoeba used in locomotion and feeding.

R

Resource partitioning – A method of reducing competition between species whereby the species utilize different regions of the habitat.

Rotifers – A small (± 100 microns) group of animals, typically with a ring of cilia at their anterior end used for locomotion and feeding.

S

Secchi disc – A disc typically about 20 cm in diameter that contains alternate quadrants of white and black. It is used to measure water transparency.

Seiche – The movement of water in a lake caused by wind, during which the water sloshes back and forth. A seiche can be in the epilimnion (a surface seiche) or the hypolimnion (an internal seiche).

Speculum – The colored patch on the wings of some birds, especially ducks.

Stoneflies – A group of animals having an extended aquatic immature stage in their life cycle. Their aquatic nymphs typically have two tail-like structures and thoracic gills. Their adult stage lasts a shorter time than their aquatic stage, and the adult's wings are folded flat on its back at rest.

Stratified lake – A lake that has formed a thermocline, separating the warmer epilimnion at the top from the cooler hypolimnion below.

T

Terminal moraine – The pile of rubble that remains after the front of an advancing lobe of a glacier retreats.

Thermocline – That portion of a lake where the water temperature changes at a rate of 1 °C or more per meter depth. This creates a thermal resistance to mixing that isolates the epilimnion from the hypolimnion.

Top-down control – The control of a give trophic level in a food web by manipulating a trophic level above it. For instance, controlling zooplankton by decreasing the population of planktivorous fish.

Trophic levels – Refers to the different strata of feeding in a food web. For instance, autotrophs are the producers that make food by photosynthesis. Herbivores are the next trophic level that eat the producers. Primary carnivores eat the herbivores, and so on.

Trophic status – Refers to the nutrient levels found in a lake. Eutrophic lakes are nutrient rich, oligotrophic lakes are nutrient poor, and mesotrophic lakes are in between.

Tundra – The most northerly vegetation zone that contains grasses, shrubs and very few stunted trees. It often has permanently frozen soil.

Turnover – The period in the seasonal cycle of a lake when the water is isothermal and wind can readily mix the entire water column. In temperate lakes, it occurs in the spring and fall.

W

Watershed – Usually considered to be the area surrounding a lake or stream that drains into the water body.

Wetlands – Areas with standing water at least part of the year. They are characterized by both soil characteristics and vegetation.

Z

Zooplankton – Weak swimming aquatic animals.

Zygote – The result of the joining of an egg and sperm. It is the first cell of a future embryo.

ENDNOTES

Chapter 1
1. Yeats, WB. 1893. The Lake Ilse of Innesfree. In Yeats. 2000. Magpie Books. Robinson Publ. Ltd. London.
2. Thoreau, HD. 1845. Walden. In Krutch, JW., ed. Walden and Other Writings by Henry David Thoreau. 1981. Bantam Books, Classic Ed. 436 pp., p 243.
3. Hamilton County Soil and Water Conservation District (HCSWCD). 2005. The State of Hamilton County Lakes: A Statistical Analysis of Water Quality 1993-2003. Publ. by HCSWCD. Lake Pleasant, NY.

Chapter 2
1. Heumann, MJ, et al. 2006. Timing of anatexis in metapelites from Adirondack lowlands and southern highlands. Geo. Soc. Am. Bull. 118: 1238-1298.
2. Bradford, BVD. 1985. Roadside Geology of New York. Mountain Press Publ. Missoula MT. p. 21.
3. Bradford. 1985. p. 27.
4. Isachsen, YW, E Landing, LV Rickard, and WB Rogers, eds. 1991. The Geology of New York. NY State Mus./Geology Survey Edu. Leaflet No. 28. State Edu. Dept. Albany, NY. 284 pp., p.38.
5. Garver, J, 2010. Geology Dept., Union College. Pers. Comm.
6. Braun, EL. 1950. The Deciduous Forests of Eastern North America. The Free Press. MacMillan Publ. Co. 596 pp.
7. Jackson, ST, and DR Whitehead. 1991. Holocene vegetation patterns in the Adirondack Mts. Ecol. Monogr. 72: 641-653.

8. Overpeck, JT. 1985. A pollen study of a late Quaternary peat bog, south-central Adirondack Mts. NY. Geol. Soc. Am. Bull. 96: 145-154.
9. Whitehead, DR, and ST Jackson. 1989. The developmental history of Adirondack (NY) lakes. J. Paleolimnology 2: 185-206.
10. Cyranoski, D. 2005. The long range forecast. Nature 438: 275-276.
11. Ritchie, WA. 1958. An Introduction to Hudson Valley Prehistory. NY State Mus. Sci Serv. Bull. 367.
12. Schneider, P. 1997. The Adirondacks: A History of America's First Wilderness. Henry Holt. Publ. 368 pp., p. 19.
13. Hochschild, HC. 1952. Township 34. A History, with Digressions, of an Adirondack Township in Hamilton County in the State of NY. Publ. Privately. 614 pp., p. 40.
14. Aber, T and S King. 1965. A History of Hamilton County. Willard Press. Booneville. 1209 pp., p. 19.
15. Aber, T and S King. 1961. Tales from an Adirondack County. Prospect Books. 208 pp., p. 4.
16. Donaldson, AL. 1921. A History of the Adirondacks, NY. v. 2 Century. 363 pp., p. 129.
17. Aber and King. 1965. p. 615.
18. Railroad Travel Brochure. 1906. Summer outings in the Sacandaga Valley. Publ. by Passenger Dept., Fonda, Johnstown and Gloversville Railroads. Found in Adiron. Mus. Lib. Blue Mt. Lake.
19. Stanyon, MP. 1965. The Quiet Years. Publ. Personally. 112 pp., p. 53.
20. Aber and King. 1965. p. 692.
21. Osborne Inn Brochure. Undated. Found in the Adiron. Mus. Lib., Blue Mt. Lake.
22. Railroad Travel Brochure. 1906.
23. Aber and King. 1965. p. 695.
24. Stanyon. 1965. p. 75.
25. Aber and King. 1965. p. 679.
26. Dunn, R. 2002. Adventures Around the Great Sacandaga Lake. Nicholas Burns Publ. Utica, NY. 170 pp., p.6.
27. Beauchamp, WM. 1907. Aboriginal place names of New York. NY State Museum Bull. 108. Archeology. p. 88.
28. Aber and King. 1961. p. 180.
29. Aber and King. 1965. p. 712.
30. Aber and King. 1961. p. 105.
31. Aber and King. 1965. p. 219.
32. Aber and King. 1965. p. 708.
33. Aber and King. 1965. p. 679.
34. Fire Protection Map of the Adirondacks. 1916. Found in Adiron. Mus. Lib. Blue Mt. Lake.
35. Fire Protection Map. 1916.
36. McMartin, B. 1994. The Great Forests of the Adirondacks. No. Country Books. 240 pp., p. 194.

Chapter 3
1. Cold Regions Research and Engineering Lab. Hanover, NH. Safety on Floating Ice Sheets. Retrieved 12 Jan 2000 (http://www.crel.usace.army.mil/ierd/ice_safety/safety.html).
2. Pielou, EC. 1998. Fresh Water. Univ. Chicago Pr. 275 pp., p. 186.
3. Pielou. 1998. p. 190.
4. Thoreau, HD. 1845. Walden. In Krutch, JW. ed. Walden and other Writings by Henry David Thoreau. 1981. Bantam Books, Classic Ed. 436 pp., p. 328.
5. Wetzel, RG. 1983. Limnology. 2nd ed. Saunders College Publ. Harcourt Brace College Publishers. 767 pp., p. 75.
6. Pielou. 1998. p. 166.
7. Pielou. 1998. p. 167.
8. McCormick, M, and G Fahnenstiel. 2002. Hydrodynamic studies on Lake Champlain. Great Lakes Env. Res. Lab. NOAA. Retrieved 12 Jan 2007 (http://www.glerl.noaa.gov/res/Task_repts/1990/ppsaylor02-1.html).
9. Wetzel. 1983. p. 56.
10. Scheffer, M. 1998. The Ecology of Shallow Lakes. Chapman and Hall. 357 pp., p. 23.
11. Larson, D. 1972. Temperature, transparency and phytoplankton productivity in Crater Lake Oregon. Limnol. and Oceanogr. 17: 410-417.

Chapter 4
1. Hamilton County Soil and Water Conservation District (HCSWCD). 2005. The State of Hamilton County Lakes: A Statistical Analysis of Water Quality 1993-2003. Publ. by HCSWCD. Lake Pleasant, NY.
2. State of New York Conservation Dept. 1933. A Biological Survey of the Upper Hudson Watershed. Biological Survey (1932) No. VII. JB Lyon Co. Printers. Albany, NY. 341 pp.
3. Bronmark, C, and LA Hansson. 2005. The Biology of Lakes and Ponds. 2nd ed. Oxford Univ. Press. 285 pp., p. 44.
4. Bronmark and Hansson. 2005. p. 47.
5. Cedar Eden Environmental. 2005. Upper Saranac Lake Research Report.
6. Jenkins, J, and A Keal. 2004. The Adirondack Atlas. Syracuse Univ. Press and the Adirondack Museum. 275 pp., p. 11.
7. McMahon, RF. 1996. The physiological ecology of the zebra mussel, *Dreisena polymorpha*, in North America and Europe. Am. Zool. 36(3): 339-363., p. 352.
8. Cohen, A. 2004. Calcium requirements and the spread of zebra mussels. Coastal Ocean Research. San Francisco Estuary Institute. Paper no. PPInvSpo4-01.
9. Schlesinger, WH. 1997. Biogeochemistry: An Analysis of Global Change. 2nd ed. Academic Press. 588 pp., p. 55.
10. Schlesinger. 1997. p. 331.
11. Jenkins, J, and A Keal. p. 243.
12. Scheuhammer, AM, et al. 1998. Mercury and selenium accumulation in common loons (*Gava immer*) and common mergansers (*Mergus merganser*) from eastern Canada. Env. Toxicol. And Chem. 17(2): 197-201.

13. NYS Federation of Lake Associations with NYS Dept. of Env. Conserv. 2009. Diet for a small lake. 2nd ed. 208 pp., p. 83.

Chapter 5
1. Goldman, CR. 1988. Primary productivity, nutrients, and transparency during the early onset of eutrophication in ultra-oligotrophic Lake Tahoe, California-Nevada. Limnol. and Oceanogr. 33(6): 1321-1333.
2. Carson, R. 1962. Silent Spring. Houghton Mifflin Press. 368 pp.
3. Woodwell, G, et al. 1967. DDT residues in an east coast estuary. A case of biological concentration of a persistent insecticide. Science 156: 821-824.

Chapter 6
1. Campbell, NA, and JB Reece. 2005. Biology. 2nd ed. Benjamin Cummings. 1247 pp., p. 521.
2. Campbell and Reece. 2005. p. 522.
3. Chiras, DD. 2001. Environmental Science. 6th ed. Jones and Bartlett. 730 pp., p. 691.
4. Campbell and Reece. 2005. p. 523.
5. Lampert, W, and U Sommer. 1997. Limnoecology: The Ecology of Lakes and Streams. Oxford Univ. Press. 382 pp., p. 67.
6. Wetzel, RG. 1983. Limnology. 2nd ed. Saunders College Publ. Harcourt Brace College Publishers. 767 pp., p. 358.
7. Mitsui, A, et al. 1986. Strategy by which nitrogen fixing unicellular cyanobacteria grow photoautotrophically. Nature 323: 720-722.
8. Sili, C, A Ena, R Materassi, and M Vincenzini. 1994. Germination of desiccated aged akinetes of alkaphilic cyanobacteria. Arch. Microbiol. 162: 20-25.
9. Roelfs, TD, and RT Oglesby. 1970. Ecological observations on the planktonic cyanophyte *Gleotrichia echinulata*. Limnol. And Oceanogr. 15: 224-229.
10. Stewart, I, et al. 2006. Cutaneous hypersensitivity reactions to freshwater cyanobacteria—human volunteer studies. BMC Dermatology 6: 6-21.
11. Li, HP, GC Gong, and TM Hsiung. 2002. Phytoplankton pigment analysis by HPLC and its application in algal community investigations. Bot. Bull. Acad. Sin. 43: 283-290.
12. Lehman, JT. 1976. Ecological and nutritional studies on *Dinobryon* Ehrenb. Seasonal productivity and the phosphate toxicity problems. Limnol. And Oceanogr. 21: 646-658.
13. Kamjunke, N, T Henrichs, and U Gaedke. 2007. Phosphorus gain by bacterivory promotes the mixotrophic flagellate *Dinobryon* spp. during re-oligotrophication. J. Plankton Res. 29: 39-46.
14. Sigee, DC, V Krivtsov, and EG Bellinger. 1998. Elemental concentrations and ratios in micro populations of *Ceratium hirundinella* (Pyrrophyta): an Xray microanalytic study. European J. of Phycol. 33: 155-164.
15. Harris, GP, SI Heaney, and JF Talling. 1979. Physiological and environmental constraints in the ecology of the planktonic dinoflagellate *Ceratium hirundinella*. Freshwater Biol. 9: 413-428.
16. Padisak, J. 1985. Population dynamics of the freshwater dinoflagellate *Ceratium hi-

rundinella in the largest freshwater lake of Central Europe, Lake Balaton, Hungary. Freshwater Biol. 15: 43-52.
17. Reynolds, CS. 1984. The Ecology of Freshwater Phytoplankton. Cambridge Univ, Press, 384 pp., p, 99.
18. Wetzel. 1983. p. 455.
19. Barker, R. 1997. And the Waters Turned to Blood. Simon and Schuster. 362 pp.
20. Sheldon, R, and CW Boylen. 1977. Maximum depth inhabited by aquatic vascular plants. Am. Midl. Natur. 97: 248-254.
21. Bradbury, J. 2004. Nature's Nanotechnologists: Unveiling the secrets of diatoms. PLoS Biol 2/10/2004: e306. Retrieved 10 Feb 2004 (http://dx.doi.org/10.1371/journal.pbio.0020306).

Chapter 7
1. Hutchinson, GE. 1975. A Treatise on Limnology. Vol. 3 Limnological Botany. John Wiley and Sons. 660 pp., p. 139.
2. Dacey, JWH. 1980. Internal winds in water lilies: An adaptation for life in anaerobic sediments. Science 210: 1017-1019.
3. Dacey, JWH. 1981. Pressurized ventilation in the yellow water lily. Ecology 65: 1137-1147.
4. Keddy, PA. 2000. Wetland Ecology. Cambridge Univ. Press. 614 pp., p. 40.
5. Barko, JW, and RM Smart. 1980. Mobilization of sediment phosphorus by submerged freshwater plants. Freshwater Biol. 10: 229-238.
6. Hutchinson. 1975. p. 146.
7. Friday, LE. 1991. The size and shape of traps in *Utricularia vulgaris* L. Functional Ecology 5: 602-607.
8. Friday, LE, and C Quarmby. 1994. Uptake and translocation of prey-derived 15N and 32P in *Utricularia vulgaris*. New Phytol. 126: 273-281.
9. Borman, S, R Korth, and J Tente. 2001. Through the looking glass: A guide to aquatic plants. Wisconsin Dept. Nat. Res. Publ. No. FH-207-97. 248 pp.
10. Richards, J. 2001. Bladder functions in *Utricularia purpurea* (Lentibulariaceae): Is carnivory important? Am. J. Bot. 88(1): 170-176.
11. Hutchinson. 1975. p. 148.
12. Wium-Andersen, S. 1971. Photosynthetic uptake of free CO_2 by the roots of *Lobelia dortmanna*. Physiol. Plant. 25: 245-248.
13. Sheldon. R, and C Boynton. 1977. Maximum depth inhabited by aquatic vascular plants. Am. Midl. Nat. 97: 248-254.
14. Borman, et al. 1997. p. 200.
15. Dorn, NJ, G Cronin, and DM Lodge. 2001. Feeding preferences and performance of an aquatic lepidopteran on macrophytes: Plant hosts as food and habitat. Oecologia 128: 406-415.
16. Keddy. 2000. p. 3.
17. Keddy. 2000. p. 59.

Chapter 8

1. Thorp, JH, and AP Covich (Ed.). 1991. Ecology and Classification of North American Freshwater Invertebrates. Academic Press. 911 pp., p. 37.
2. Hutchinson, GE. 1967. A treatise on Limnology. Vol. 2. Introduction to Lake Ecology and the Limnoplankton. John Wiley and Sons. 1115 pp., p. 491.
3. Edmondson, WT (Ed.) 1959. Freshwater Biology. 2nd ed. John Wiley and Sons. 1248 pp., p. 426.
4. Thorp and Covitch. 1991. p. 188.
5. Thorp and Covitch. 1991. p. 197.
6. Ruttner-Kolisko, A. 1974. Zooplankton Rotifers: Biology and Taxonomy. Die Binnengewasser. Vol. XXVI/1 Supplement. 145 pp., p. 33.
7. Bennet, WN, and ME Boraas. 1989. A demographic profile of the fastest growing metazoan: a strain of *Brachionus calyciflorus* (Rotifera). Oikos 55: 365-360.
8. Thorp and Covitch. 1991. p. 208.
9. Starkweather, PL. 1980. Aspects of rotifer feeding behavior. Hydrobiologia 73: 63-72.
10. Stemberger, RS, and JJ Gilbert. 1987. Multiple-species induction of morphological defenses in the rotifer *Keratella testudo*. Ecology 68(2): 370-378.
11. Gilbert, JJ. 1966. Rotifer ecology and embryological induction. Science 151: 1234-1237.
12. Thorp and Covitch. 1991. p. 21.
13. Thorp and Covitch. 1991. p. 731.
14. Gilbert, JJ. 1985. Competition between rotifers and *Daphnia*. Ecology 66: 1943-1950.
15. Engle, DL. 1985. The production of haemoglobin by small pond *Daphnia pulex*: intraspecific variation and its relation to habitat. Freshwater Biology 15: 631-638.
16. Dini, ML, and SR Carpenter. 1991. The effect of whole-lake fish community manipulations on *Daphnia* migratory behavior. Limnol. and Oceanogr. 36(2): 370-377.
17. Thorp and Covitch. 1991. p. 790.
18. Makarewicz, JC, and GE Likens. 1979. Structure and function of the zooplankton community of Mirror Lake, New Hampshire. Ecol. Monogr. 49: 109-127.
19. Willamson, CE, and NM Butler. 1986. Predation on rotifers by the suspension-feeding calanoid copepod *Diaptomis pallidus*. Limnol. and Oceanogr. 31(2): 393-402.
20. McCafferty, WP. 1981. Aquatic Entomology. Jones and Bartlett. Boston. 448 pp., p. 70.
21. Sweeney, BW, and RL Vannote. 1982. Population synchrony in mayflies: a predator satiation hypothesis. Evolution 36(2): 810-821., p. 811.
22. Fremling, CR. 1964. Mayfly distribution indicates water quality on the upper reaches of the Mississippi River. Science 14: 1164-1166.
23. Flecker, AS. 1992. Fish predation and the evolution of inveretbrate drift periodicity: evidence from a neotropical stream. Ecology 73 (2): 438-448.
24. Sweeney and Vannote. 1992. p. 820.
25. Craven, RE, and BE Brown. 1969. Ecology of Hexagenia niads (Insects-Ephemeroptera) in an Oklahoma reservoir. Am. Midl. Nat. 82 (2): 346-358.

26. McCafferty. 1981. p. 150.
27. Wiggins, GB. 1996. Larvae of the North American Caddisfly Genera (Trichoptera). 2nd ed. Univ. Toronto Press. 457 pp., p.15.
28. McCafferty. 1981. p. 239.
29. Tjossem, SF. 1990. Effects of fish chemical cues on vertical migration behavior of Chaoborus. Limnol. and Oceanogr. 35(7): 1456-1468.
30. McCafferty. 1981. p. 303.

Chapter 9
1. Freeman, S. 2005. Biological Science. 2nd ed. Pearson/Benjamin Cumming. 1283 pp., p. 767.
2. Campbell, NA, and JB Reece. 2005. Biology. 7th ed. 1230 pp., p. 683.
3. Smith, CL. 1985. The Inland Fishes of New York State. NYS Dept. of Env. Conserv. Albany, NY. 522 pp., p. 312.
4. Lampert, W, and U Sommer. 1997. Limnoecology. Oxford Univ. Press. 382 pp., p. 217.
5. Brown, GE, and S Brennan. 2000. Chemical alarm signals in juvenile green sunfish (*Lepomis cyanellus*, Centrachidae). Copeia(4): 1079-1082.
6. Smith, CL. 1985.
7. Scott, WB, and EJ Crossman. 1973. Freshwater Fishes of Canada. Bull. 184. Fisheries Board of Canada. 966 pp., p. 513.
8. Sibley, DA. 2000. The Sibley Guide to Birds. Nat. Audubon Soc. Alfred Knopf. 544 pp.
9. Palmer, RS. 1976. Handbook of North American Birds. Vol. 2. Yale Univ. Press. 521 pp., p. 321.
10. Palmer, RS. 1976. Handbook of North American Birds. Vol. 3. Yale Univ. Press. 560 pp., p. 488.
11. Palmer, RS. 1988. Handbook of North American Birds. Vol. 4. Yale Univ. Press. 433 pp., p. 94.
12. Palmer. 1988. p. 97.
13. Palmer, RS. 1976. Handbook of North American Birds. Vol. 1. Yale Univ. Press. 567 pp., p. 35.
14. Wisconsin Dept. Natural Resources. 2003. Mercury and loons: A wildlife risk assessment. Aquiring Knowledge. Wisconsin DNR Bureau of Integrated Science Services. Biennial Rept. 2003. pp. 6-7.
15. Cahalane, VH. 1961. Mammals of North America. MacMillan Co. 682 pp., p. 188.
16. Cahalane. 1961. p. 452.
17. White, RS. 1999. Animal Skulls. Internat. Wildlife Mus. Tuscon, AZ. 53 pp., p. 9.
18. White. 1999. p. 7.

Chapter 10
1. Lack, D. 1947. Darwin's Finches. Cambridge Univ. Press. 208 pp., p. 62.
2. Grant, PR, and BR Grant. 2006. Evolution of character displacement in Darwin's finches. Science 313(5784): 224-226.
3. MacArthur, RH. 1958. Population ecology of some warblers of northeastern coniferous

forests. Ecology 39(4): 599-619.
4. Futuyma, DJ. 1986. Evolutionary Biology. Sinauer Assoc. 600 pp.
5. Hutchinson, GE. 1958. Homage to Santa Rosalia or why are there so many kinds of animals? Am. Natur. 93: 145-159.
6. Edmondson, WT, and JT Lehman. 1981. The effect of changes in nutrient income on the condition of Lake Washington. Limnol. and Oceanogr. 26(1): 1-29.
7. Wetzel, RG. 1983. Limnology. 2nd ed. Saunders College Publishing. 767 pp., p. 375.
8. Wetzel. 1983. p. 377.
9. Brooks, JL, and SI Dodson. 1965. Predation, body size and composition of plankton. Science 150: 28-35.
10. Kitchell, JF (ed). 1992. Food Web Management: A Case Study of Lake Mendota. Springer-Verlag. 550 pp.
11. Johnson, BM, and MD Staggs. 1992. The fishery. pp. 354-375. In: Kitchell, JF (ed). Food Web Management: A Case Study of Lake Mendota. Springer-Verlag. 550 pp.
12. Kitchell, JF, and SR Carpenter. 1992. Summary: Accomplishments and new directions of food web management in Lake Mendota. pp. 539-544. In: Kitchell, JF (ed). Food Web Management: A Case Study of Lake Mendota. Springer-Verlag. 550 pp.
13. Lathrop, RC, SR Carpenter, and DM Robertson. 1999. Summer water clarity responses to phosphorus, *Daphnia* grazing, and internal mixing in Lake Mendota. Limnol. and Oceanogr. 44(1): 137-146.
14. Otsego County Conservation Association (OCCA). 15 Jul 2010. Otsego Lake Water Quality Data.
15. Carpenter, SR, et al. 2001. Trophic cascades, nutrients, and lake productivity: Whole-lake experiments. Ecological Monogr. 71(2): 163-186.

Chapter 11

1. Jenkins, J, K Roy, C Driscoll, and C Buerkell. 2007. Acid Rain in the Adirondacks: An Environmental History. Comstock Publ. Cornell Univ. Press. 246 pp., p. 35.
2. Schofield, CL. 1976. Acid precipitation: Effects on fish. Ambio. 5: 228-230.
3. Jenkins, et al. 2007. p. 75.
4. Howells, G. 1995. Acid Rain and Acid Waters (2nd ed.). Ellis Horwood Press. 262 pp., p. 142.
5. Hendrey, GR (ed.). 1984. Early Biotic Responses to Advancing Lake Acidification. Acid Precipitation Series. Vol. 6. Ann Arbor Science Book. Butterworths Publ. 173 pp.
6. Johnson, DW, HA Simonin, JR Colquhoun, and FM Flack. 1987. *In situ* toxicity tests of fish in acid waters. Biogeochemistry 3: 181-208.
7. Baker, JP, WJ Warren-Hicks, J Gallagher, and SW Christensen. 1993. Fish population losses from Adirondack lakes: The role of surface water acidity and acidification. Water Resources Research 29: 861-874.
8. Jenkins, et al. 2007. p. 195.
9. US Envioronmental Protection Agency (EPA). 2009. Acid Rain and Related Programs: 2007 Progress Report. Doc. EPA-430-K-08-010. 46 pp., p. 1.
10. Driscoll, CT, KM Driscoll, KM Roy, and MJ Mitchell. 2003. Chemical response of lakes in the Adirondack region of New York to declines in acidic deposition. Environ. Sci. Technol. 37: 2036-2042.

11. Driscoll, et al. 2003.
12. Jenkins et al. 2007. p. 199.
13. Driscoll, CT, et al. 2007. Mercury contamination in forest and freshwater ecosystems in the northeastern United States. Bioscience 57: 17-28.
14. Driscoll, et al. 2007.
15. Driscoll, et al. 2007.
16. NYS Dept. of Env. Conserv. 2008. Fishing License booklet.
17. Chen, CY, RS Stemberger, NC Kamman, BM Mayes, and CL Folt. 2005. Pattern of Hg bioaccumulation and transfer in aquatic food webs across multi-lake studies in the northeastern US. Ecotoxicology 14: 135-147.
18. Evers, DC, et al. 2008. Adverse effects from environmental mercury loads on breeding common loons. Ecotoxicology 17: 69-81.
19. Schoch, N, and DC Evers. 2002. Monitoring Mercury in Common Loons: New York Field Report, 1998-2000. Report BRI 2000-1 submitted to US Fish Wildl. Serv. and NYS Dept. of Env. Conserv. Biodiversity Research Institute, Falmouth, ME.
20. Jenkins, J, and A Keal. 2004. The Adirondack Atlas. Syracuse Univ. Press and Adirondack Museum. 275 pp., p. 242.
21. Driscoll, et al. 2007.
22. Kamman, NC, and DR Engstrom. 2002. Historical and present fluxes of mercury to Vermont and New Hampshire lakes inferred from ^{210}Pb dated sediment cores. Atmospheric Environment 36: 1599-1609.
23. Evers, et al. 2008.
24. Aquatic Control Technology, Inc. 2006. Draft Environmental Impact Statement for Saratoga Lake Invasive Species Long-Term Management.
25. Albany Times Union. 29 Jun 2009. Keeping weeds at bay in Saratoga Lake.
26. Lake Colby Association. 2009. Invasives. Retrieved 9 Mar 2009 (http://www.lakecolby.org/invasives/invasives.htm).
27. Upper Saranac Lake Association. 2009. Retrieved 9 Mar 2009 (http://www.uslf.org/milfoil).
28. Mascoma Lake Association. 2009. Retrieved 9 Mar 2009 (http://www.mascomalakeassociation.org).
29. Adirondack Park Invasive Plant Program. 2010. Annual Report 2010.
30. US National Park Service. 2009. Retrieved 9 Mar 2009 (http://www.nps.gov/plants/alien/fact/lysa1).
31. Malecki, RA, et al. 1993. Biological control of purple loosestrife. Bioscience 43: 680-686.
32. Albright, MF, et al. 2004. Recovery of native flora and behavioral responses by *Galerucella* sp. following biocontrol of purple loosestrife. Am. Midl. Nat. 52: 248-254.
33. Strayer, DL. 1999. Effects of alien species on freshwater mollusks in North America. Journal of the North American Benthlogical Society. 18: 74-98.
34. Strayer, DL, et al. 1999. Transformation of freshwater ecosystems by bivalves: A case study of zebra mussels in the Hudson River. Bioscience 49: 19-27.
35. Cohen, A. 2004. Calcium requirements and the spread of zebra mussels. Coastal Ocean Research. San Francisco Estuary Institute. Paper no. PPInvSpo4-01.

36. US Geological Survey. 2009. Retrieved 9 Mar 2009 (http://www.nas.usgs.gov/queries/Factsheet.asp?speciesID=162).
37. Straile, D, and A Halbich. 2000. Life history and multiple antipredator defenses of an invertebrate pelagic predator, *Bythotrepes longimanus*. Ecology 81: 150-163.
38. Preall, R. 2010. NYS Dept. of Env. Conserv., Fisheries Biologist, Region 5. Pers. Comm.
39. Henderson, CL, CJ Dindorf, and FJ Rozumalski. 2009. Landscaping for Wildlife and Water Quality. Minnesota Dept. of Natural Resources. 176 pp., p. 37.
40. Smith, CL. 1985. The Inland Fishes of NY State. NYS Dept. of Env. Conserv. 522 pp.
41. State of New York Conservation Dept. 1933. A Biological Survey of the Upper Hudson Watershed. Biological Survey (1932) No. VII. JB Lyon Co. Printers. Albany, NY. 341 pp., p. 112, 189.

Chapter 12

1. Christensen, JH, et al. 2007. Regional Climate Projections. In: Climate Change 2007: The Physical Basis. Contributions of Working Group 1 to the Fourth Assessment Report of the Intergovernmental Panel on Climate Change. Cambridge Univ. Press.
2. Union of Concerned Scientists. 2007. Confronting Climate Change in the US Northeast: Impacts and Solutions. A Report of the Northeast Climate Change Impacts Assessment 2007. 8 pp.
3. UN Meteorological Organization, 2011. Press release no. 934. 21 Nov 2011.
4. Oechel, WC, et al. 1993. Recent change of Artic tundra ecosystems from a carbon sink to a source. Nature 361: 520-523.
5. Kiehl, J. 2011. Lessons from the Earth's past. Science 331: 1858-159.
6. Audubon Society. 2009. Bird movements reveal global warming threat in action. Press release. 10 Feb 2009.
7. Cornell University Project Feeder Watch. 2009. Retrieved 10 Feb 2009 (http://www.birds.cornell.edu/pfw/).
8. Union of Concerned Scientists. 2007.
9. Henderson, CL, CJ Dindorf, and FJ Rozumalski. 2009. Landscaping for Wildlife and Water Quality. Minn. Dept. Nat. Res. 176 pp.
10. Colborn, T, D Dumanoski, and JP Meyers. 1997. Our Stolen Future. Plume, Penguin Books. 316 pp.

Appendix III

1. Pielou, EC. 1998. Fresh Water. Univ. Chicago Pr. 275 pp., p. 156.

Acknowledgements

I AM DEEPLY INDEBTED TO the following people and organizations for their contributions to the creation of this book. Without their help, the book would have remained only a dream.

I would first like to thank my colleague in the Biology Department at Union College, Carl George, who cajoled me into joining the EPA-funded project to study the dredging of Collins Lake shortly after I arrived at Union. This project introduced me to the finer points of lake ecology and redirected my research interests toward aquatic biology. My colleagues in the Geology Department at Union—Kurt Hollocher, John Garver and Don Rodbell—assisted me with water chemical analysis and discussions regarding the geology of the Adirondacks. And Tom Werner of the Chemistry Department helped with various topics concerning water chemistry. Union College provided me with research facilities and other support while I worked on this book.

Then there are my students, far too many to mention here, who have helped gather the data on nearby aquatic systems mentioned in this book. But I will specifically mention Kim Maison, who helped with some experiments conducted at Sacandaga Lake, and Ian Brennan, Sarah Coleman, Lisa Crescenzo, Cristi DeBernardo, and Kim Maison, who all contributed fish illustrations to the book from our Illustrated Organism class.

The following organizations were also instrumental in the preparation of the book. Research libraries at the Adirondack Museum at Blue Mountain Lake and at the Association for the Protection of the Adirondacks Center in Schenectady were very helpful in my search for arcane Adirondack information. The Citizen's Statewide Lake Assessment Program (CSLAP) sponsored by the New York State Department of Environmental Conservation provided valuable water quality data.

Once a manuscript emerged, the following people read it in various iterations and provided valuable guidance: Kathy Armstrong, Jere Brophy, Willard Harman, Chet Harvey, Scott Kishbaugh, Rick Mincher, John Rasmussen, Nancy and Glen Slack, Emily Southgate, Bill Thielking and Jon Tobiessen. In addition, the participants in an Academy for Lifelong Learning (Empire State College) class in Saratoga Springs during the fall of 2010 read a draft and provided useful verbal and written feedback. Especially helpful were comments by Gail Rheingold, Carolee DeBlaere, Eric Krantz and Claire Olds.

I also thank the editors at Graphite Press for taking a chance and publishing this book.

—P. T.